METAL IONS IN
BIOLOGICAL SYSTEMS

VOLUME 20

Concepts on Metal Ion Toxicity

METAL IONS IN BIOLOGICAL SYSTEMS

Edited by

Helmut Sigel
Institute of Inorganic Chemistry
University of Basel
Basel, Switzerland

with the assistance of Astrid Sigel

VOLUME 20
Concepts on Metal Ion Toxicity

MARCEL DEKKER, INC.   New York and Basel

Library of Congress Catalog Number: 79-640972

MARCEL DEKKER, INC.
270 Madison Avenue, New York, New York 10016

ISSN: 0161-5149
ISBN: 0-8247-7540-6

Current printing (last digit):
10 9 8 7 6 5 4 3 2 1

PRINTED IN THE UNITED STATES OF AMERICA

Preface to the Series

Recently, the importance of metal ions to the vital functions
of living organisms, hence their health and well-being, has become
increasingly apparent. As a result, the long-neglected field of
"bioinorganic chemistry" is now developing at a rapid pace. The
research centers on the synthesis, stability, formation, structure,
and reactivity of biological metal ion-containing compounds of low
and high molecular weight. The metabolism and transport of metal
ions and their complexes is being studied, and new models for complicated natural structures and processes are being devised and
tested. The focal point of our attention is the connection between
the chemistry of metal ions and their role for life.

   No doubt, we are only at the brink of this process. Thus,
it is with the intention of linking coordination chemistry and
biochemistry in their widest sense that the series METAL IONS IN
BIOLOGICAL SYSTEMS reflects the growing field of "bioinorganic
chemistry." We hope, also, that this series will help to break
down the barriers between the historically separate spheres of
chemistry, biochemistry, biology, medicine, and physics, with the
expectation that a good deal of the future outstanding discoveries
will be made in the interdisciplinary areas of science.

   Should this series prove a stimulus for new activities in
this fascinating "field," it would well serve its purpose and would
be a satisfactory result for the efforts spent by the authors.

Fall 1973   Helmut Sigel

Preface to Volume 20

In Volume 18 of this series the *Circulation of Metals in the Environment* was considered. The discussion of the relation between the chemistry of the elements and their biogeochemical cycles reveals the double role metal ions are playing in the physiology of organisms: some are indispensable for normal life, while most of them are toxic at elevated concentrations, that is, they harmfully affect the activity of living organisms. Indeed, recent years have brought an increasing concern for the potential toxic effects of metal ions, which constitute part of the products and byproducts of our technologies. It is not possible to cover all the related aspects of the subject in a single book; therefore this volume focuses the *Concepts on Metal Ion Toxicity* with the aim of improving our understanding of the relation between the chemistry of metal ions and their toxicological properties.

First the distribution of potentially hazardous trace metals in the atmosphere, hydrosphere, and lithosphere is considered, and the bioinorganic chemistry of metal ion toxicity is discussed. The interrelations between essentiality and toxicity of metals and between metal ion speciation and toxicity in aquatic systems are outlined. Next, metal toxicity to agricultural crops, to humans and animals, including nutritional aspects, is described. Other chapters deal with the chromosome damage in individuals exposed to heavy metals and the mechanistic aspects of metal ion carcinogenesis. The volume closes by describing methods for the in vitro assessment of metal ion toxicity and by summarizing problems encountered in the analysis of biological materials for toxic trace elements.

Helmut Sigel

Contents

CONTENTS

Chapter 6

METAL ION TOXICITY IN MAN AND ANIMALS

Paul B. Hammond and Ernest C. Foulkes

Chapter 7

HUMAN NUTRITION AND METAL ION TOXICITY

M. R. Spivey Fox and Richard M. Jacobs

Chapter 8

CHROMOSOME DAMAGE IN INDIVIDUALS EXPOSED TO HEAVY METALS

Alain Léonard

Contributors

Numbers in parentheses indicate the pages on which the authors'
contributions begin.

Frank T. Bingham Department of Soil and Environmental Sciences,
     University of California, Riverside, California 92521 (119)

Max Costa* Department of Pharmacology, University of Texas,
     Medical School, Houston, Texas 77225 (259/279)

Elie Eichenberger Federal Institute for Water Resources and Water
     Pollution Control, Department of Technical Biology, EAWAG's
     Experiment Station, Tüffenwies 36, CH-8064 Zurich, Switzerland
     (67)

Ernest C. Foulkes Institute of Environmental Health, Kettering
     Laboratory, University of Cincinnati College of Medicine,
     Cincinatti, Ohio 45267-0056 (157)

M. R. Spivey Fox Division of Nutrition, Food and Drug Administra-
     tion, Department of Health and Human Services, Washington, D.C.
     20204 (201)

Paul B. Hammond Institute of Environmental Health, Kettering
     Laboratory, University of Cincinnati College of Medicine,
     Cincinatti, Ohio 45267-0056 (157)

J. Daniel Heck Lorillard Research Center, Greensboro, North Carolina
     27420 (259/279)

Richard M. Jacobs Division of Nutrition, Food and Drug Administra-
     tion, Department of Health and Human Services, Washington, D.C.
     20204 (201)

Wesley M. Jarrell Department of Soil and Environmental Sciences,
     University of California, Riverside, California 92521 (119)

---

*Present affiliation: Department of Environmental Medicine and
Pharmacology, New York University Medical Center, New York City,
New York 10016

Alain Léonard Mammalian Genetics Laboratory, Department of Radio-
    biology, Center for the Study of Nuclear Energy, B-2400 Mol,
    Belgium, and Teratogenicity and Mutagenicity Unit, University
    of Louvain, Brussels, Belgium (229)

R. Bruce Martin Chemistry Department, University of Virginia,
    Charlottesville, Virginia 22901 (21)

Gordon K. Pagenkopf Department of Chemistry, Montana State
    University, Bozeman, Montana 59717 (101)

Frank J. Peryea* Department of Soil and Environmental Sciences,
    University of California, Riverside, California 92521 (119)

Hans G. Seiler Institute of Inorganic Chemistry, University of
    Basel, CH-4056 Basel, Switzerland (305)

Garrison Sposito Department of Soil and Environmental Sciences,
    University of California, Riverside, California 92521 (1)

---

*Present affiliation: Tree Fruit Research Center, Washington State
University, Wenatchee, Washington 99163

Contents of Other Volumes

---

*Out of print

———————————

*Out of print

Other volumes are in preparation.

Comments and suggestions with regard to contents, topics, and the like for future volumes of the series would be greatly welcome.

METAL IONS IN BIOLOGICAL SYSTEMS

VOLUME 20

Concepts on Metal Ion Toxicity

1

Distribution of Potentially Hazardous Trace Metals

Garrison Sposito
Department of Soil and Environmental Sciences
University of California
Riverside, California 92521

1.   INTRODUCTION

1.1.   Potentially Hazardous Trace Metals

Environmental geochemistry is concerned with the distribution of
chemical elements among the atmosphere, hydrosphere, and lithosphere of the earth as it bears on problems of environmental pollution [1]. Of special interest to this relatively new discipline

are the metal elements in the periodic table whose abundance in the
hydrosphere and atmosphere has been increased significantly by the
accelerated extraction of minerals and fossil fuels from the lithosphere and by other technological processes characteristic of the
present century. Most of these metal elements were to be found
typically at total concentrations well below 1 mmol/m^3 in pristine
fresh and saline waters of the hydrosphere. For this reason they
are referred to as "trace metals." Rapid, localized increases of
trace metal concentrations in the hydrosphere or atmosphere commonly
are observed concomitantly to the development of exploitative technologies. This kind of sudden change exposes the local biosphere to
the risk of destabilization, thereby eliciting a concern for public
health and an institutional classification of the perturbing metals
as "potentially hazardous." The extent to which a particular trace
metal actually poses an environmental hazard depends not only on the
rate and magnitude of its enrichment in the hydrosphere or atmosphere,
but also on its chemical speciation and the details of its biochemical cycling [1-4].

A list of potentially hazardous trace metals is provided in
Table 1. The selection is not exhaustive, even to the extent of
representing all of the environmentally important oxidation states
of any one metal, but it is likely to include most of the candidate
elements slated for regulatory action on the basis of potential
biosphere impact [4-7]. Besides the metals themselves, three important geochemical parameters related to atomic structure also are
listed in Table 1. Crystallographic ionic radii based on the authoritative compilation by Shannon and Prewitt [8,9] appear in the second
column. Because these ionic radii are empirical constructs based on
measured interatomic distances in crystals and assumed values for
the radii of O(-II) and F(-I) [8], their relevance to the (unsolvated)
radii of metal cations in aqueous solutions--the biological milieu--
can be questioned. Fortunately, enough information about the molecular structure of electrolyte solutions is available to lay this question to rest permanently. Shown in Table 2 are nearest-neighbor

TABLE 1

Fundamental Atomic Structural Properties of
Some Potentially Hazardous Trace Metals

Trace metal	Ionic radius[a] (nm)	Ionic potential (nm^{-1})	Misono parameter (nm)
Li(I)	0.074	13.5	0.053
Be(II)	0.027	74.1	0.032
Al(III)	0.053	56.6	0.073
V(IV)	0.059	67.8	0.211
Cr(III)	0.062	48.4	0.226
Mn(II)	0.083	24.1	0.273
Fe(II)	0.078	25.6	0.291
Co(II)	0.075	26.7	0.270
Ni(II)	0.069	29.0	0.252
Cu(II)	0.073	27.4	0.284
Zn(II)	0.075	26.7	0.240
Sr(II)	0.113	17.7	0.202
Mo(VI)	0.042	142.9	0.092
Ag(I)	0.115	8.7	0.405
Cd(II)	0.095	21.1	0.303
Sn(IV)	0.069	58.0	0.194
Sb(III)	0.077	39.0	0.254
Cs(I)	0.170	5.9	0.264
Hg(II)	0.102	19.6	0.396
Pb(II)	0.118	16.9	0.393
U(VI)	0.073	82.2	—
Pu(III)	0.100	30.0	—

[a]Abstracted from data compiled by Shannon and Prewitt [8,9].

ion-water molecule distances (d) compiled recently by Marcus [10]
from published x-ray or neutron diffraction studies and Monte Carlo
or molecular dynamics computer experiments on aqueous solutions.
These data can be compared to the corresponding Shannon-Prewitt

Table 2

Nearest-Neighbor Ion-Water Molecule Distances (d) and
Crystallographic Ionic Radii (R) of 23 Elements

Ion	d^a (nm)	R^b (nm)	Ion	d^a (nm)	R^b (nm)
Li(I)	0.207	0.074	Co(II)	0.209	0.075
F(-I)	0.274	0.133	Ni(II)	0.207	0.069
Na(I)	0.237	0.102	Cu(II)	0.198	0.073
Mg(II)	0.211	0.072	Zn(II)	0.208	0.075
Al(III)	0.188	0.053	Br(-I)	0.333	0.196
Cl(-I)	0.312	0.181	Cd(II)	0.234	0.095
K(I)	0.273	0.138	I(-I)	0.361	0.220
Ca(II)	0.242	0.100	Cs(I)	0.308	0.170
Cr(III)	0.199	0.062	La(III)	0.251	0.106
Mn(II)	0.220	0.083	Nd(III)	0.248	0.098
Fe(III)	0.204	0.073	Gd(III)	0.239	0.094
Fe(II)	0.212	0.078	Lu(III)	0.234	0.086

[a]Data compiled by Marcus [10].
[b]Data compiled by Shannon and Prewitt [8,9].

crystallographic radii (R), which are also listed in Table 2.   If
the crystallographic radii are the same as the unsolvated ionic
radii in aqueous solution, then a linear relationship between d and
R should be found in which the slope has unit value and the y intercept equals the equivalent spherical radius of a water molecule.
Regression analysis of the data in Table 2 in fact produced the
linear expression

$$d = 0.136 \pm 0.006 + 1.01 \pm 0.05 \ R \qquad (r^2 = 0.9859^{**}) \qquad (1)$$

where the deviations represent confidence limits at P = 0.05.   The
slope in Eq. (1) has the value expected and the intercept value of
0.136 ± 0.006 nm is a respectable estimate of the radius of a water
molecule [11].

The third column in Table 1 lists the ionic potential IP = Z/R,
where Z is the oxidation number of a metal.   This parameter can be

used to classify metal ions as to the nature of their interactions
with nearest-neighbor water molecules in dilute aqueous solutions
at equilibrium with atmospheric CO_2 (pH 5.6) [12]. If $Z/R < 30$, a
metal cation will be only solvated by water molecules; if $95 > Z/R >
30$, a metal cation can repel protons strongly enough from a solvating
water molecule to form a hydrolytic species; if $Z/R > 95$, the repul-
sion is strong enough to allow the formation of oxyion species.
Typical examples of trace metals placed in these three categories
are Ag (8.7), Al (56.6), and Mo (142.9), with their ionic potential
given in parentheses.

The fourth column in Table 1 lists values of the Misono softness parameter Y [13]:

$$Y = \frac{10 I_Z R}{I_{Z+1} \sqrt{Z}} \tag{2}$$

where I_Z is the ionization potential (not the ionic potential) of
a metal with oxidation number Z [14] and R is its ionic radius in
nanometers. The Misono parameter provides a quantitative measure
of the tendency of a metal cation to be polarizable and enter into
covalent bonding [15,16]. Other parametric measures of this tendency
have been proposed by Nieboer and Richardson [17] and Williams [18].
Like these other proposed indices of covalency, Y can be used to give
quantitative significance to the classification of metal cations as
"hard" or "class a," "borderline," and "soft" or "class b" Lewis acids
[15-19]. Hard Lewis acids are molecular units characterized by high
charge density, high electronegativity, and low polarizability. They
tend to form strong complexes stabilized by large entropy increases
and primarily electrostatic interactions. Soft Lewis acids, on the
other hand, are characterized by low charge density, low electronegativity, and high polarizability. They tend to form strong
complexes stabilized by large enthalpy decreases and covalent interactions [15].

Figure 1 illustrates the hard-soft Lewis acid classification
for the trace metals listed in Table 1. Hard acids correspond to
$Y < 0.25$, borderline acids to $0.25 < Y < 0.32$, and soft acids to

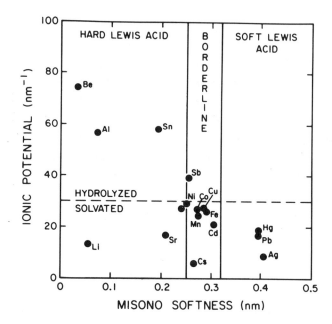

FIG. 1. Classification of the trace metals in Table 1 according to their hydrolyzable and Lewis acid characters.

Y > 0.32 (Y in nanometers). These divisions match closely those in the class a-borderline-class b classification proposed by Ahrland et al. [20]. It is noteworthy in Fig. 1 that the first-row transition metals, as well as Cu(II) and Cd(II), fall into the borderline category; that is, they can behave as either hard or soft acids, depending on solvent, stereochemical, and electronic structural factors. Although it is not entirely evident from Fig. 1, macro-nutrient metals [e.g., Na(I) and Ca(II)] tend to be hard acids, whereas micronutrient metals [e.g., Fe(II) and Cu(II)] tend to be borderline acids. Representative biochemical properties of class a, borderline, and class b metal ions have been summarized by Nieboer and Richardson [17].

1.2. Metal Ion Toxicity and Atomic Structure

The ionic potential is a useful atomic structural parameter for characterizing the natural abundance of elements in plants and animals. This perspective, introduced by Hutchinson [21] and elaborated by Banin and Navrot [22], is illustrated in Fig. 2 through log-log plots of an enrichment factor EF_B:

$$EF_B = \frac{\text{average metal concentration in organism}}{\text{average metal concentration in earth's crust}} \qquad (3)$$

versus ionic potential IP for terrestrial higher plants and animals [22]. The shapes of the two plots are remarkably like one another and like the same kind of graph for bacteria and for fungi [22]. The enrichment factor appears to show a shallow minimum around IP $\approx$ 20 nm^{-1}, then a precipitous drop to a deep minimum near IP $\approx$ 60 nm^{-1}, and then finally a gradual rise toward very large

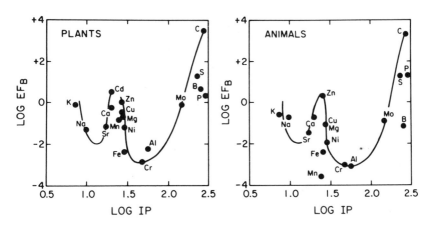

FIG. 2. Banin-Navrot plots for chemical elements found in terrestrial higher plants and animals. Enrichment factor data compiled by Banin and Navrot [22].

values as IP increases above 100 nm^{-1}. The sharp drop in EF_B, by
almost three orders of magnitude, occurs very near IP = 30 nm^{-1},
the "hydrolysis edge" (Fig. 1), and evidently reflects the rapidly
decreasing solubility of hydrolyzable metals in pristine natural
waters as their ionic potentials increase. The broad depression
centered on IP $\approx$ 60 nm^{-1} encompasses ionic potentials in the range
of 20-150 nm^{-1}, and this range may be taken to define a set of
metals whose natural abundance in living organisms is low. Large
anthropogenic increases in the concentrations of these metals in
the hydrosphere or atmosphere are fundamentally inimical to the
evolution of a biosphere in which they are relatively depleted.
In this respect, Fig. 2 and the IP range between 20 and 150 nm^{-1}
provide a simple geochemical criterion for determining which
chemical elements are likely to be "potentially hazardous."

A refinement of this picture of metal hazard can be obtained
by considering the molar toxicity sequences compiled by Nieboer and
Richardson [17], summarized in Table 3. For each kind of organism,
the ordering of the metals in Table 3 from left to right reflects
an increasing molar quantity of metal required to produce substantial toxic effect, with the smallest molar quantity required being

TABLE 3

Representative Molar Toxicity Sequences[a]

Organisms	Toxicity sequence
Algae	Hg > Cu > Cd > Fe > Cr > Zn > Co > Mn
Fungi	Ag > Hg > Cu > Cd > Cr > Ni > Pb > Co > Zn > Fe
Flowering plants	Hg > Pb > Cu > Cd > Cr > Ni > Zn
Annelids	Hg > Cu > Zn > Pb > Cd
Fish	Ag > Hg > Cu > Pb > Cd > Al > Zn > Ni > Cr > Co > Mn >> Sr
Mammals	Ag,Hg,Cd > Cu,Pb,Co,Sn,Be >> Mn,Zn,Ni,Fe,Cr >> Sr > Cs,Li,Al

[a]Abstracted from data compiled by Nieboer and Richardson [17].

associated with the metal judged most toxic in the sequence. Once again there is a remarkable similarity among organisms. Reference to Fig. 1 indicates that, in general, *soft metal cations are more toxic than borderline metal cations, which are more toxic than hard metal cations.* Since covalent bonding is the hallmark of the soft Lewis acid, metal toxicity must be enhanced by the possibility of forming strong complexes with biomolecules [17].

2. DISTRIBUTION OF TRACE METALS IN THE ATMOSPHERE, HYDROSPHERE, AND LITHOSPHERE

2.1. Atmospheric Concentrations

Table 4 lists representative concentrations of the potentially hazardous trace metals in Table 1 for the vicinal air over Europe and North America. Most of the data represent median atmospheric

TABLE 4

Atmospheric Concentrations of Some Potentially Hazardous Trace Metals over Europe and North America (ng/m^3 at STP)[a]

Metal	Europe	North America	Metal	Europe	North America
Li	2	—	Sr	2	3
Be	2	0.2	Mo	1	2
Al	600	1500	Ag	1	1
V	30	400	Cd	20	21
Cr	25	60	Sn	20	20
Mn	43	150	Sb	8	12
Co	2	3	Cs	0.5	0.2
Ni	25	100	Hg	1	2
Cu	340	100	Pb	120	2700
Zn	1200	500	U	0.02	0.2

[a]Most of the data represent median atmospheric concentrations selected from the compilations by Bowen [5] and Kabata-Pendias and Pendias [7].

concentrations selected from the compilations by Bowen [5] and
Kabata-Pendias and Pendias [7]. It is well known that trace metal
concentrations in air show geographical variations reflecting technological development [5,7,23,24]. The data in Table 4 should be
interpreted to represent the effect of urban--but not highly
industrialized--regions on trace metal concentrations [23]. It is
also well known that atmospheric concentrations are very difficult
to measure and that pronounced variability exists among the data
published for a particular metal. This important point is illustrated in Fig. 3 through a comparative plot of the ranges of atmospheric trace metal concentrations over North America reported by
Bowen [5] and Kabata-Pendias and Pendias [7]. The general agreement
between the two compilations is good, but substantial differences as
to the range of concentration are exemplified by Al, Cu, and Pb.
This kind of discrepancy implies that concentration differences

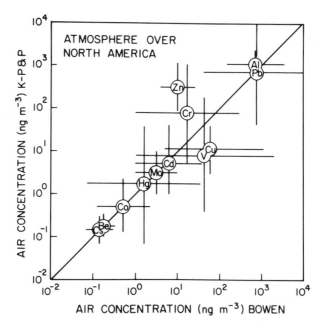

FIG. 3. Comparison of atmospheric concentration data compiled by
Kabata-Pendias and Pendias [7] with those compiled by Bowen [5].
Lines indicate the ranges of compiled values.

between Europe and North America probably should not be interpreted
as significant unless they represent an order of magnitude or more.
Among the metals in Table 4, irrespective of continent, the concentrations of Al, Cu, Zn, Pb, and perhaps V stand out as large in an
absolute sense. Since the flux of an atmospheric particle toward
the earth is equal to the product of its concentration and its
deposition velocity, these prominent concentrations suggest relatively large fluxes of the metals to the soil surface.

2.2. Hydrospheric Concentrations

Table 5 lists median concentrations of the trace metals in Table 1
for freshwater and seawater, taken directly from the compilation by
Bowen [5]. The caveat pertaining to the range of values for a particular metal and to differences among compilations, given above,
should be remembered here as well. Bowen [5] has suggested that

TABLE 5

Freshwater and Seawater Concentrations of Some
Potentially Hazardous Trace Metals $(mg/m^3)^a$

Metal	Fresh	Saline	Metal	Fresh	Saline
Li	2	180	Sr	70	8000
Be	0.3	0.006	Mo	0.5	10
Al	300	2	Ag	0.3	0.04
V	0.5	2.5	Cd	0.1	0.1
Cr	1	0.3	Sn	0.009	0.004
Mn	8	0.2	Sb	0.2	0.2
Co	0.2	0.02	Cs	0.02	0.3
Ni	0.5	0.6	Hg	0.1	0.03
Cu	3	0.3	Pb	3	0.03
Zn	15	5	U	0.4	3

[a]Data abstracted from the compilation by Bowen [5].

elements for which the ratio of freshwater to seawater concentration
exceeds 100 are to be regarded as showing significant enrichment in
freshwaters relative to seawater (principally rivers compared to
oceans). According to this criterion, the data in Table 5 indicate
that the metals Al, probably Pb, and possibly Be are so enriched
(as is Fe [5]). These trace metals, accordingly, are predicted to
be found in large concentrations in solid detritus at the mouths of
rivers. Bowen [5] also suggests that those elements which are
enriched in seawater relative to freshwater (Li, V, Sr, Mo, Cs, and
U in Table 5) will cycle more slowly than the water in the ocean.

2.3. Lithospheric Concentrations

Table 6 lists median soil and crustal concentrations of the trace
metals in Table 1, also taken directly from compilations by Bowen
[5]. The concentration ranges and median values for these metals

TABLE 6

Soil and Crustal Concentrations of Some Potentially
Hazardous Trace Metals (mg/kg)[a]

Metal	Soil	Crust	Metal	Soil	Crust
Li	25	20	Sr	250	370
Be	0.3	3	Mo	1	2
Al	71,000	82,000	Ag	0.05	0.07
V	90	160	Cd	0.4	0.1
Cr	70	100	Sn	4	2
Mn	1,000	950	Sb	1	0.2
Co	8	20	Cs	4	3
Ni	50	80	Hg	0.06	0.05
Cu	30	50	Pb	35	14
Zn	90	75	U	2	2

[a]Abstracted from data compiled by Bowen [5].

are similar to those reported independently by Brooks [25] and
Kabata-Pendias and Pendias [7], but comparative plots of the data
would resemble Fig. 3.   Kabata-Pendias and Pendias [7] have made an
exhaustive survey of trace metal concentrations in soils and rocks.
Their detailed compilation shows the variability to be expected in
these concentrations according to geographic region, soil or rock
type, and extent of contamination.   In nonpolluted regions there is
a close correspondence between soil and crustal concentrations of
trace metals.   This characteristic, illustrated in Fig. 4 for the
trace metals in Table 1, is an effect of the relatively low mobility
of the metals during natural pedogenesis.   Since the primary chemical
agent in pedogenesis is the proton, anthropogenic increases in soil
proton input through acid deposition can be expected to accelerate
pedogenesis and alter the relationship depicted in Fig. 4.   The same

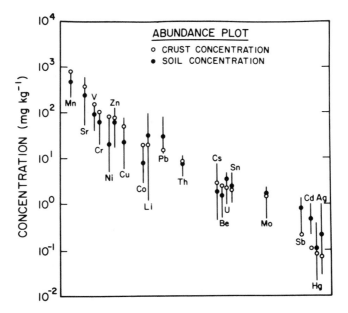

FIG. 4.   Mean concentrations and ranges of trace metals in topsoils
compared with mean crustal concentrations, based on data compiled
by Kabata-Pendias and Pendias [7].

cannot be said for the effect of plants grown in soil; the depletion of trace metals by nonaccumulating plants appears to be negligible [23].

3. METAL ENRICHMENT AND TRANSFER

3.1. Metal Enrichment Factors

Fluid pathways for the global transfer of hazardous trace metals are provided by the atmosphere and by large rivers which empty into the oceans. The land mass, the beds of rivers, and the oceans act as (temporary) reservoirs for the metals. The extent to which the atmosphere is supplying metals to either the land or the ocean surface in excess of natural cycling inputs can be assessed through an examination of the enrichment factor EF_A:

$$EF_A = \frac{J_M}{J_{IM}} \cdot \frac{\text{surface concentration of IM}}{\text{surface concentration of M}} \tag{4}$$

where J_M is the average flux of metal M to the land or ocean surface and J_{IM} is the same kind of flux for an index metal assumed to have a negligible anthropogenic throughput in the atmosphere. Common choices of IM include Al, Sc, Ti, and Fe [23]. Aluminum is a convenient choice because of the size and accuracy of its data base.

Table 7 lists values of EF_A based on IM = Al for the transfer of the metals in Table 1 to the ocean surface near Europe (North Sea) and near North America (tropical North Atlantic Ocean). Data on atmospheric fluxes reported by Buat-Ménard [26] were substituted into Eq. (3) along with ocean concentrations of the metals taken from Table 5. (The median ocean concentration of Th reported by Bowen [5] was used along with flux data to calculate EF_A for this metal because no flux value for U was available.) The small magnitude of EF_A in Table 5, with the notable exception of Pb, can be interpreted as evidence that the ocean receives little or no anthropogenic excess deposition of trace metals from the atmosphere. The independence of

TABLE 7

Enrichment Factors (Atmosphere/Sea) for Some
Potentially Hazardous Trace Metals[a]

Metal	North Sea	North Atlantic	Metal	North Sea	North Atlantic
Al	1.0[b]	1.0[b]	Zn	0.12	0.010
V	0.013	0.0003	Ag	—	0.009
Cr	0.047	0.019	Cd	0.029	0.02
Mn	0.31	0.14	Sb	0.019	0.0007
Co	0.13	0.054	Hg	—	0.028
Ni	0.029	0.013	Pb	5.9	4.1
Cu	0.29	0.033	Th	0.27	0.36

[a]Calculations based on atmospheric fluxes compiled by Buat-Ménard [26].
[b]By definition.

this conclusion from the choice of index metal is supported by EF_A values for Sc and Fe based on IM = Al, the flux data of Buat-Ménard [26], and the concentration data of Bowen [5]. For the North Sea EF_A is 0.9 for Fe and 0.6 for Sc; for the tropical North Atlantic, EF_A is 0.6 for Fe and 0.7 for Sc. These values are not significantly different from 1.0 assumed for Al.

Table 8 lists values of EF_A for the transfer of the metals in Table 1 to the land surface of Europe and North America, based on flux data reported by Sposito and Page [23] and the soil concentration data in Table 6. Significant anthropogenic deposition ($EF_A >$ 100) is indicated for Cu, Zn, Ag, Cd, Sn, Sb, Hg, Pb, and possibly Be, V, and Mo. The values for Ag, Cd, and Hg are particularly high.

Anthropogenic transfer of trace metals through rivers can be evaluated with an enrichment factor EF_W:

$$EF_W = \frac{M_{FW}}{Al_{FW}} \cdot \frac{Al_S}{M_S} \tag{5}$$

where M_{FW} is the median concentration of a metal in freshwater (Table 5) and M_S is its median concentration in soil (Table 6).

TABLE 8

Enrichment Factors (Atmosphere/Soil) for Some
Potentially Hazardous Trace Metals

Metal	Europe	North America	Metal	Europe	North America
Li	9.5	—	Sr	1.0	0.6
Be	789	32	Mo	118	95
Al	1.0^a	1.0^a	Ag	2367	947
V	39	210	Cd	5917	2485
Cr	42	41	Sn	592	237
Mn	5.1	7.1	Sb	947	568
Co	30	18	Cs	15	2.4
Ni	59	95	Hg	1972	1578
Cu	1341	158	Pb	406	3651
Zn	1578	263	U	1.2	4.7

[a]By definition.

Equation (5) is not a completely adequate estimator because it does
not incorporate trace metal transport in the sediment load of rivers
and is not weighted according to metal solubility. Table 9 lists
values of EF_{FW} that suggest important anthropogenic transport of Ag
and Hg in river systems.

TABLE 9

Enrichment Factors (Freshwater/Soil) for Some
Potentially Hazardous Trace Metals

Metal	EF_{FW}	Metal	EF_{FW}	Metal	EF_{FW}
Li	19	Co	5.9	Cd	59
Be	237	Ni	2.4	Sn	0.5
Al	1.0^a	Cu	24	Sb	47
V	1.3	Zn	39	Cs	1.2
Cr	3.4	Sr	66	Hg	394
Mn	1.9	Mo	118	Pb	20
Fe	3.0	Ag	1420	U	47

[a]By definition.

3.2. Metal Transfer Rates

The metal transport scheme implicit in the foregoing discussion of
enrichment factors is diagrammed in Fig. 5. Tables 7 and 8 provided
information concerning the transfer processes AO (air-to-ocean) and
AL (air-to-land) under the assumption that Al cycles through the
atmosphere without significant anthropogenic perturbation [23,26,27].
Data pertaining to the rates of the transfer processes LA, OA, LR,
and RO in Fig. 5 are given in Table 10. The rate of the transfer
process LA includes both anthropogenic and natural emissions [26],
the latter representing soil, volcanoes, and vegetation as sources.
The predominance of the LA rate over the OA rate is parallel to the
dominance of AL deposition over AO deposition already noted. Anthropogenic contributions to LA are the more important ones for Mn, Cu,
Zn, Cd, and Pb [26].

The fourth column in Table 10 gives estimates of the natural
contribution to the rate of the transfer process LR reported by
Bowen [5]. These data can be compared to the mining rates listed
in the sixth column of Table 10 to provide yet another criterion for
designating a trace metal "potentially hazardous" [5]. If the mining
rate exceeds LR_{na} by more than an order of magnitude, the metal involved is a candidate for regulation. On this basis, one concludes

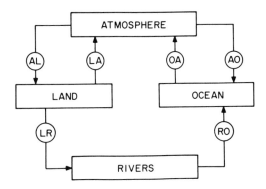

FIG. 5.   Diagram of the global cycling of trace metals, with rect-
angles indicating storage components and circles indicating transfer
processes.

TABLE 10

Mass Transfer Rates for Some Potentially
Hazardous Trace Metals (10^6 kg/yr)

Metal	LA[a]	OA[a]	LR$_{na}$[b]	RO[c]	Mining[c]
Al	$2.5 \cdot 10^4$	150	$3.3 \cdot 10^5$	$1.1 \cdot 10^4$	$1.3 \cdot 10^5$
V	53	20	640	19	18
Mn	622	5	$3.8 \cdot 10^3$	300	$2.5 \cdot 10^4$
Fe	$1.6 \cdot 10^4$	50	$1.6 \cdot 10^5$	$1.9 \cdot 10^4$	$6.8 \cdot 10^5$
Cu	69	4	200	110	$6.2 \cdot 10^3$
Zn	380	5	300	550	$4 \cdot 10^3$
Cd	5	0.5	0.4	3.7	7.7
Pb	408	5	56	110	$3.3 \cdot 10^3$

[a]Data compiled by Buat-Ménard [26].
[b]Natural (na) transfer by weathering; data compiled by Bowen [5].
[c]Data compiled by Bowen [5]; RO includes dissolved load only.

that Cu, Zn, Cd, and Pb are potentially hazardous trace metals.
Similar reasoning adds Cr, Ag, Sn, Sb, and Ag to the list [5].

The difference between the rate of the transfer process RO
and that of LR$_{na}$ provides a minimum estimate of the anthropogenic
contribution to RO when this difference is positive. The estimate
is a minimum because the RO data do not include sediment transport
and because some anthropogenic contributions to LR may remain in
untransported sediment within the river system or be volatilized.
Even with this qualification, the anthropogenic contribution to LR
for Zn and Pb is significant and for Cd it is spectacular.

Despite many uncertainties concerning the verisimilitude of
the concentration and flux data, the conclusion emerges from the
present review that the trace metals Cu, Zn, Ag, Cd, Sb, Sn, Hg,
and Pb are the most potentially hazardous on a global or regional
scale. Lead is of acute concern on the global scale because of its
prominent showing in all of the enrichment factors and transfer
rates considered. These conclusions echo those reached by Andreae

et al. [28] in a recent Dahlem Group Report. The significant enrichment of the natural environment in Pb, Hg, Cd, and Ag, paired with the low natural abundance of these metals in the biosphere [5] and their high degree of Lewis acid softness in aqueous milieux (Fig. 1), indicate a persistent, special need for study and monitoring in relation to public health.

ACKNOWLEDGMENTS

Gratitude is expressed to Dr. J. P. LeClaire for statistical analyses of the data in Table 2 and to Ms. Louise DeHayes for typing the manuscript of this review.

REFERENCES

1. I. Thornton, *Applied Environmental Geochemistry,* Academic Press, London, 1983.

2. S. H. U. Bowie and J. S. Webb, *Environmental Geochemistry and Health,* The Royal Society, London, 1980.

3. C. J. M. Kramer and J. C. Duinker, *Complexation of Trace Metals in Natural Waters,* Martinus Nijhoff/Dr. W. Junk, The Hague, The Netherlands, 1984.

4. J. O. Nriagu, *Changing Metal Cycles and Human Health,* Springer-Verlag, Berlin, 1984.

5. H. J. M. Bowen, *Environmental Chemistry of the Elements,* Academic Press, London, 1979.

6. J. M. Wood and E. D. Goldberg, in *Global Chemical Cycles and Their Alterations by Man* (W. Stumm, ed.), Dahlem Konferenzen, Berlin, 1977, p. 137.

7. A. Kabata-Pendias and H. Pendias, *Trace Elements in Soils and Plants,* CRC Press, Boca Raton, Florida, 1984.

8. R. D. Shannon and C. T. Prewitt, *Acta Crystallogr. Sect. B, 25,* 925 (1969).

9. R. D. Shannon and C. T. Prewitt, *Acta Crystallogr. Sect. B, 26,* 1046 (1970).

10. Y. Marcus, *J. Solution Chem., 12,* 271 (1983).

11. G. Sposito, *J. Chem. Phys.*, *74*, 6943 (1981).

12. D. Carroll, *Rock Weathering*, Plenum Press, New York, 1970.

13. M. Misono, E. Ochiai, Y. Saito, and Y. Yoneda, *J. Inorg. Nucl. Chem.*, *29*, 2685 (1967).

14. L. H. Ahrens, *Ionization Potentials*, Pergamon Press, Oxford, 1983.

15. G. Sposito, *The Thermodynamics of Soil Solutions*, Oxford University Press, Oxford, 1981.

16. J. E. Huheey, *Inorganic Chemistry*, Harper and Row, New York, 1972.

17. E. Nieboer and D. H. S. Richardson, *Environ. Pollut. Ser. B*, *1*, 3 (1980).

18. R. J. P. Williams, in *Changing Metal Cycles and Human Health* (J. O. Nriagu, ed.), Springer-Verlag, Berlin, 1984, p. 251.

19. W. B. Jensen, *The Lewis Acid-Base Concepts: An Overview*, John Wiley, New York, 1980.

20. S. Ahrland, J. Chatt, and N. R. Davies, *Q. Rev. Chem. Soc.*, *12*, 265 (1958).

21. G. E. Hutchinson, *Q. Rev. Biol.*, *18*, 1 (1943).

22. A. Banin and J. Navrot, *Science*, *189*, 550 (1975).

23. G. Sposito and A. L. Page, in *Metal Ions in Biological Systems*, Vol. 18 (H. Sigel, ed.), Marcel Dekker, New York, 1984, p. 287.

24. W. H. Zoller, in *Changing Metal Cycles and Human Health* (J. O. Nriagu, ed.), Springer-Verlag, Berlin, 1984, p. 27.

25. R. R. Brooks, in *Environmental Chemistry* (J. O'M. Bockris, ed.), Plenum Press, New York, 1977, p. 429.

26. P. E. Buat-Ménard, in *Changing Metal Cycles and Human Health* (J. O. Nriagu, ed.), Springer-Verlag, Berlin, 1984, p. 43.

27. D. H. Peirson and P. A. Cawse, *Philos. Trans. R. Soc. London*, *B288*, 41 (1979).

28. M. O. Andreae, T. Asami, K. K. Bertine, P. E. Buat-Ménard, R. A. Duce, Z. Filip, U. Förstner, E. D. Goldberg, H. Heinrichs, A. B. Jernelöv, J. M. Pacyna, I. Thornton, H. J. Tobschall, and W. H. Zoller, in *Changing Metal Cycles and Human Health* (J. O. Nriagu, ed.), Springer-Verlag, Berlin, 1984, p. 359.

2

Bioinorganic Chemistry of Metal Ion Toxicity

R. Bruce Martin
Chemistry Department
University of Virginia
Charlottesville, Virginia 22901

1. INTRODUCTION

There are five behavior categories for metal ions and other substances in living systems. An *essential nutrient* is one for which a deficiency results in impairment of function that is relieved only by administration of that substance. Essentiality is organism dependent, and care must be taken to distinguish it from stimulation. There are many instances of both essential and nonessential metal ions behaving as *stimulants*. Some metals and metal ions at some concentrations may be merely inert or *innocuous*, without evident effect. Inert metals such as tantalum, platinum, silver, and gold are often employed as surgical implants. There are numerous metal ions that serve as *therapeutic agents*. Examples include the use of mercury compounds against parasites, a zinc carboxylate against the fungi that causes "athlete's foot," and lithium in manic depression [1]. Finally, at high concentrations most metal ions become *toxic*: They harmfully affect the activity of living organisms, perhaps irreversibly, in a manner that leads to loss of function, deformity, and perhaps death. Depending upon concentration and time, a single

substance may act in more than one of the five ways in a single
organism.

2. ESSENTIALITY AND TOXICITY

Figure 1 schematizes the biological response of a tissue to increasing concentration of an essential nutrient. The solid curve is
deliberately drawn with an immediate positive response above zero
concentration as if the essential nutrient is saturating a site and
no nutrient is consumed at the lowest levels by other interactions.
(Other interactions are, of course, possible.) The solid curve
levels off at an optimal level that spans a wide concentration
range for many essential metal ions. Finally, at excess concentrations deleterious effects begin to set in, the biological response

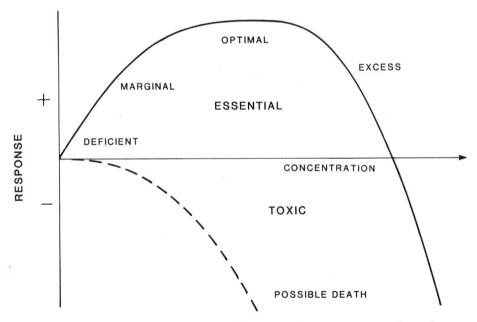

FIG. 1. Biological response dependence on tissue concentration of
an essential nutrient (solid curve) and of a deleterious substance
(dashed curve). The relative position of the two curves on the
concentration axis is arbitrary and one of convenience.

becomes unfavorable, and the substance becomes toxic. The dashed
curve in Fig. 1 illustrates the biological response to a wholly
harmful substance without any essential or stimulating effects.
The dashed curve is drawn with a lag region under the assumption
that an organism can cope with some amounts of a substance (threshold) before toxic effects become evident. Figure 1 shows general
characteristics; each substance has its own specific curve of biological response versus concentration.

Figure 1 emphasizes the point that an essential nutrient
may become toxic at sufficiently high concentrations. (Chapter 3
develops this interrelationship for plants.) Almost any substance
in excess ultimately becomes harmful, even if the action is indirect, as in limiting the absorption of other essential nutrients.
Animals maintain a concentration within the optimal range by a complex set of physiological reactions termed homeostasis. Concentrations of all essential metal ions are under homeostatic control.
The detailed mechanism of homeostasis for many metal ions remains
an area of current research.

A current list of metal ions essential to humans (and mammals)
appears in Table 1. As experimentation has continued and techniques
have become more refined, some metal ions once considered to be only
toxic have been found to be essential. Ni^{2+} has not yet been proved
as essential in man. Some other metals such as tin have been suggested as essential in mammals [2].

The second column in Table 1 lists the form the metal ion takes
up at pH 7 and that would occur in the plasma unless intercepted by
other ligands. The FeO(OH) and CuO solids do not occur in plasma,
as both Fe^{3+} and Cu^{2+} are complexed by proteins. The third column
in Table 1 lists a typical amount of each essential trace element
found in a normal adult [3]. The corresponding plasma concentration
appears in the fourth column. The last column of Table 1 lists a
recommended adult daily allowance for each essential metal ion.
Many of these recommendations are subject to change.

In response to foreign matter, organisms possess detoxification
mechanisms that serve to limit or even eliminate a toxic substance.

TABLE 1

Essential Metal Ions

Metal ion	pH 7 form	Human amount	Plasma concentration	Daily allowance
Na^+	Na^+	100 g	141 mM	1-3 g
K^+	K^+	140 g	4 mM	2-5 g
Mg^{2+}	Mg^{2+}	25 g	0.9 mM	0.7 g
Ca^{2+}	Ca^{2+}	1100 g	1.3 mM	0.8 g
Cr^{3+}	$Cr(OH)_2^+$	6 mg	0.5 µM	0.1 mg
$Mo(VI)$	MoO_4^{2-}	9 mg		0.3 mg
Mn^{2+}	Mn^{2+}	12 mg	1 µM	4 mg
Fe^{3+}	$FeO(OH)\downarrow$	4-5 g	20 µM	10-20 mg
Fe^{2+}	Fe^{2+}	4-5 g	20 µM	10-20 mg
Co^{2+}	Co^{2+}	1 mg	0.5 µM	3 µg[a]
Ni^{2+}	Ni^{2+}	10 mg	0.05 µM	–[b]
Cu^{2+}	$CuO\downarrow$	0.1 g	19 µM	3 mg
Zn^{2+}	Zn^{2+}	2 g	46 µM	15 mg

[a]Of vitamin B_{12}.
[b]Essentiality indicated in animals only.

Study of specialized detoxification mechanisms for metal ions is at
an early stage. Many metal ions render themselves less damaging by
making use of their own properties, such as in the formation of
insoluble compounds in the gut, transport by the blood to other
tissues where the metal ion might be immobilized (e.g., Pb^{2+} in the
bone), or conversion by the liver and kidney to some less toxic or
disposable form. In response to a challenge from toxic Cd^{2+}, Hg^{2+},
Pb^{2+}, and other metal ions, human liver and kidney organs respond
by increasing synthesis of metallothionein, a small protein in which
approximately one-third of the 61 amino acid residues are cysteine
[4]. The high occurrence and frequent juxtaposition of sulfhydryl
groups provide strong binding sites for toxic metal ions.

The mechanisms by which metal ions become toxic are easy to imagine but difficult to pinpoint for any single metal ion. Metal ions stabilize and activate many proteins; perhaps one-third of all enzymes require metal ions. Competitions between natural and toxic metal ions for a protein site are easily conceptualized. Many proteins contain sulfhydryl groups that may become metalated by toxic heavy metal ions such as Cd^{2+}, Hg^{2+}, and Pb^{2+}, and it is widely believed that this reaction is a likely mode of toxicity of these metal ions. Nonetheless, the exact proteins at which the most consequential damage to the organism occurs remain unknown. Toxic metal ions become distributed among many tissues, and there is no guarantee that the greatest damage occurs where the metal ion is most plentiful. An example is Pb^{2+}, where more than 90% becomes immobilized in bone, but the toxic effects are rendered by the remaining 10% distributed among other tissues. Indeed, the immobilization of Pb^{2+} in bone may be viewed as a detoxifying mechanism. Our sparse knowledge of toxicity mechanisms limits the incisiveness of this review. Toxicity due to genetic diseases such as Cooley's anemia with its iron overload is not considered in this chapter.

This review does not cover possible carcinogenic effects of metal ions. Carcinogenicity is complex, depending upon animal, organ, level, and synergistic effects with other substances. For roles of individual metal ions see the compendia Refs. 3 and 5, for a general review see Volume 10 of this series [6], and for mechanistic aspects see Chap. 9 of this volume. Metal ions and their complexes may also serve as anticancer agents, a role discussed in Volume 11 of this series [7]. This review also excludes radiation damage due to metal ions.

Metal ion toxicity is usually unrelated to any essentiality. Toxicity and essentiality do share, however, one feature: a frequent interdependence among metal ions and between metal ions and nonmetals in contributing to effectiveness. The availability of essential metal ions depends upon interactions with other foods eaten; mere adequacy of diet does not suffice. For example, iron in vegetables is poorly absorbed owing to complexing anions and excess Zn^{2+} may inhibit Cu^{2+} absorption. In like manner, Cd^{2+} toxicity appears worse in a Zn^{2+}-deficient system, and Pb^{2+} toxicity more severe with a Ca^{2+} deficiency. These antagonisms and mutual dependencies complicate simple guidelines and explanations for essentiality and toxicity [8]. Other examples are given in Sec. 4.

3. PROPERTIES OF METAL IONS

3.1. Ionic Radii

Metal ions commonly substitute for others of the same size, providing a basis for metal ion toxicity. Table 2 presents effective ionic radii [9] for most of the metal ions discussed in this chapter. (Depending upon how observed bond lengths are divided among contributing atoms or ions, absolute values of radii may differ in various compilations. Whatever the basis for the division, however, radii from a single compilation exhibit constant differences upon comparison of cations with each other.) Note that for a given ion the radius increases with coordination number, which should be specified when quoting radii. The radius for the most common coordination number is underlined in Table 2. Typically the alkali and alkaline-earth metal ions exhibit variable coordination numbers without strong directionality in bonding. Table 2 shows that Co^{2+} and Zn^{2+} exhibit highly similar radii in all coordination numbers.

Size similarity usually ranks at higher importance than charge identity in permitting metal ion substitutions. Thus Ca^{2+} and Na^{+} of closely similar size but different charges often interchange with one another. For example, the plagioclase feldspars display a continuous series of aluminum silicate mixtures in a common lattice from $NaAlSi_3O_8$ to $CaAl_2Si_2O_8$. Neither Mg^{2+} nor Sr^{2+} replace Ca^{2+} in this series. The charge compensation required in minerals is not critical when substitution occurs in a protein or in other biological ligands.

TABLE 2

Effective Ionic Radii (Å)

Ion	Coordination number					
	4	5	6	7	8	9
Be^{2+}	0.27		0.45			
Al^{3+}	0.39	0.48	0.54			
Cr^{3+}			0.62			
Fe^{3+}	0.49	0.58	0.65[a]		0.78	
Ni^{2+}	0.55	0.63	0.69			
Mg^{2+}	0.57	0.66	0.72		0.89	
Cu^{2+}	0.57	0.65	0.73			
Co^{2+}	0.58	0.67	0.74		0.90	
Zn^{2+}	0.60	0.68	0.74		0.90	
Li^+	0.59		0.76		0.92	
Fe^{2+}	0.63		0.78[b]		0.92	
Mn^{2+}	0.66	0.75	0.83	0.90	0.96	
Gd^{3+}			0.94	1.00	1.05	1.11
Cd^{2+}	0.78	0.87	0.95	1.03	1.10	
Ca^{2+}			1.00	1.06	1.12	1.18
Na^+	0.99	1.00	1.02	1.12	1.18	1.24
Sr^{2+}			1.18	1.21	1.26	1.31
Pb^{2+}	0.98		1.19	1.23	1.29	1.35
Ba^{2+}			1.35	1.38	1.42	1.47
K^+	1.37		1.38	1.46	1.51	1.55
Tl^+			1.50		1.59	
Rb^+			1.52	1.56	1.61	1.63
Cs^+			1.67		1.74	1.78

[a]High spin; low-spin value is 0.55.
[b]High spin; low-spin value is 0.61.
Source: Ref. 9.

3.2. Stability Sequences

From stability constant tables [10], increasing metal ion stabilities follow these orders:

$$Ca < Mg < Mn < Fe \sim Cd \sim Pb < Co \sim Zn < Ni \ll CH_3Hg^+ \sim Cu \ll Hg$$

for glycine and

$$Mg \ll Mn \ll Fe < Cd \sim Co \sim Zn \ll Pb, Ni < CH_3Hg^+ \ll Cu \lll Hg$$

for 1,2-diaminoethane (en). The ion Ca^{2+} does not form stable
amine complexes. Except for methyl mercury (CH_3Hg^+), all the
metal ions carry two positive charges. Owing to the strongly
chelating bidentate ligands glycine and en, CH_3Hg^+ with only a
single strong coordinating site is at a competitive disadvantage
in the above series. To a unidentate ligand CH_3Hg^+ binds more
strongly than all of the above metal ions except Hg^{2+}. In the
above series each inequality sign represents an approximate 10-fold
increase in stability constant. The two series for glycine and en
are similar, the major difference being a stability constant span
of 10^8 for glycine and 10^{14} for en from Mg^{2+} to Hg^{2+}. As a generalization, the increment between metal ions increases on passing
from O < N < S donor atoms. Unfortunately, too few single ligands
with a sulfur donor have been measured with enough metal ions to
compile a comparable list. However, cross-comparing results on
several ligands and metal ions indicate that the presence of a
sulfur donor promotes Cd^{2+} and Pb^{2+} to ranks higher than their
positions in the above series. The order of increasing sulfur-
binding strengths is $Cd^{2+} \lesssim Pb^{2+} < CH_3Hg^{2+} < Hg^{2+}$. These metal ions
are especially toxic, as they are capable of disrupting the biochemistry of an organism by combining with sulfhydryl groups on both
small molecules and proteins.

3.3. Stability Ruler

For a single ligand the Irving-Williams stability sequence of
dipositive metal ions is Mg < Mn < Fe < Co < Ni < Cu > Zn, invariant
of ligand. We use the uniformly progressive part of the sequence
from Mg^{2+} to Cu^{2+} to propose a *stability ruler* against which more
variable metal ion stabilities may be compared. The stability
ruler appears across the top of Table 3. For several ligand donor
sets Table 3 shows the relative binding strengths of the variable
dipositive metal ions Zn^{2+}, Cd^{2+}, Hg^{2+}, and Pb^{2+}. Their placement
within Table 3 corresponds to their stability constants compared
with those of the metal ions on the stability ruler. The entries
for glycine and en correspond to the series in the previous section.
The variable metal ions need not be dipositive, but all are in this
case. With most of the metal ions in Table 3 histidine will be
tridentate [11,12]. The increment between metal ions and the width
of the stability ruler scale increase with the substitutions O < N <
S. Thus oxygen donor ligands are least discriminating between metal
ions, nitrogen donors intermediate, and sulfur donors most discriminating.

The stability ruler of Table 3 shows that with all donor sets
except the one involving sulfur (and OH^-; see Sec. 3.4), Zn^{2+} stabilities resemble those of Co^{2+}. When a sulfhydryl group is present,
as in 2-mercaptoethylamine, Zn^{2+} binding strengthens, becoming as
strong as Ni^{2+} in this case. For 2-mercaptoethylamine Table 3 also
shows relative strengthening of Cd^{2+} and Pb^{2+} binding to the ligand
with a sulfhydryl donor.

To most unidentate ligands Cd^{2+} binds more strongly than Zn^{2+}.
Probably because of an unfavorable small chelate ring bite size for
the relatively large Cd^{2+} (Table 2), Zn^{2+} chelates more strongly to
ligands containing O and N donors. Upon introduction of a S donor
atom into a chelate, Cd^{2+} becomes the stronger metal ion binder
(Table 3).

TABLE 3

Stability Ruler[a]

Ligand	Donors	Mg <	Mn <	Fe <	Co <	Ni <	Cu	Hg	Width[b]
Oxalate	O,O		Cd		Zn Pb			Hg (2.3)[c]	2.1
Glycine	N,O			Cd Pb	Zn			Hg (0.4)	6.1
Histidine	N,=N			Cd Pb	Zn			Hg (0.2)	8.2
$NH_2(CH_2)_2NH_2$	N,N			Cd	Zn	Pb		Hg (0.4)	10.1
$NH_2(CH_2)_2S^-$	N,S						Zn Cd Pb	Hg (0.3)	11[d]
Imidazole	=N				Zn Cd			Hg (1.1)	4.0
Ammonia	NH_3					Zn Cd		Hg (1.2)	4.0
Hydroxide	OH^-		Cd				Zn Pb	Hg (1.2)	3.7

[a]Cations all dipositive in this application.
[b]Ruler scale width in difference between log stability constants of Cu^{2+} and Mg^{2+} complexes.
[c]Relative magnitude by which the Mg to Cu log stability constant scale must be extended to reach the Hg value.
[d]Estimated owing to the reduction of Cu^{2+}.

For all donor sets in Table 3, Hg^{2+} binds so strongly that it
is off the end of the ruler scale. The number in parentheses after
Hg in Table 3 is the relative distance by which the whole log stability constant scale from Mg to Cu must be extended to reach the
value for Hg. An instructive contrast appears between the magnitudes
of the extension for most bidentate ligands compared to unidentate
ligands. Except for oxalate (which appears as a special case), the
scale extension for Hg^{2+} amounts to 0.2-0.4 log units for bidentate
ligands and to 1.1-1.2 log units for the last three unidentate
ligands in Table 3. This difference arises because Hg^{2+} prefers
linear 2-coordination and binds the second donor atom in small
chelate rings much more weakly than the first donor atom.

3.4. Metal Ion Hydrolysis

The higher the charge density or the ratio of charge to the radius,
the more likely it is for a metal ion to undergo hydrolysis in
aqueous solutions to give hydroxo complexes. Hydroxo complexes may
abruptly form polynuclear complexes and precipitates. The first
four small metal ions in Table 2 undergo hydrolysis even in acidic
solutions and form precipitates. In 6-coordination the charge-to-
radius ratio for the first four metal ions is greater than 4.4,
while for all the remaining metal ions in Table 2 the ratio is less
than 2.9. The first four metal ions cannot occur in the bloodstream
(pH 7.4) as the free aqueous ion and must be complexed in some way.
Covalency may also promote hydroxo complex formation in acidic solutions, as is the case for Hg^{2+}.

Hydroxo complexes follow the same stability order as other
ligands, with the strongest binding to the right on the stability
ruler in Table 3, where the hydroxide ion appears as the last entry.
The ion Zn^{2+} exhibits a relatively strong tendency to form hydroxo
complexes and even precipitates, creating problems in interpretations of Zn^{2+} complexing [13]. The enhanced tendency of Zn^{2+} to
form hydroxo complexes may be due to its readiness to take up

tetrahedral coordination with its smaller radius and greater charge-to-radius ratio (Table 2).

For consistency in this chapter all metal ion hydrolysis results are taken from a single authoritative source [14].

3.5. Hard and Soft Acids and Bases

Probably unfortunately for biochemical applications, the hard-soft classification of acids and bases has become part of the popular vocabulary. According to the precept, "hard" or class a metal ions favor interactions with "hard" bases, and "soft" or class b metal ions favor combinations with "soft" bases [15,16]. In biochemistry there are only four main types of ligand donor atoms of which O and aliphatic N are classified as "hard," aromatic N as "borderline," and S as "soft" bases. All alkali, alkaline-earth, and lanthanide ions are classified as hard, as are the first four metal ions in Table 2 and Mn^{2+}. Borderline metal ions include the last four in the stability ruler (Table 3) with Zn^{2+} and Pb^{2+}. Thus the first eight essential metal ions in Table 1 are hard and the last five are borderline. Soft metal ions include Cd^{2+}, CH_3Hg^+, Hg^{2+}, and Tl^+. The classification scheme is limited to three categories each for acids and bases without a quantitative scale.

Qualitatively we can see some correspondences between the hard and soft classifications and our more quantitative comparisons in the previous four sections. The first four metal ions in Table 2, all hard, form insoluble hydroxide (hard) precipitates, even in acidic solutions. However, 11 other hard metal ions in Table 2 do not form hydroxo complexes at pH 7 and many do not do so even in strongly basic solutions. Some borderline and soft metal ions form hydroxo complexes in neutral solutions, while others do not.

As mentioned in our stability ruler (Table 3) comparisons, there is a tendency for the soft S to combine with the soft metal ions, but also with Pb^{2+}, classified as borderline, but not with other borderline metal ions. In fact, Pb^{2+} defies useful classification because it interacts strongly with sulfhydryl groups (soft), and also forms an anomalously strong hydroxide (hard) complex. That a single metal ion exhibits characteristics of all three classes greatly weakens the utility of the hard and soft classification scheme.

A proposed test of "softness" is the ability to interact with an ether S such as occurs in methionine [12]. Several complexes are known where borderline metal ions do not interact with an ether S, but interactions do occur with Co^{3+} (hard), Cu^{2+} (borderline), and Pd^{2+}, Pt^{2+}, Ag^{+}, CH_3Hg^{+}, and Hg^{2+} (all soft), but not with Cd^{2+} (soft). On this basis it was suggested that Cd^{2+} behaves more like a borderline metal ion [12]. Interestingly, an attempt at quantitative evaluation also assigns Cd^{2+} to a borderline category [17].

Biology provides ligand donors to metal ions that are inconsistent with and would not be predicted from the hard-soft classification scheme. As one example, in the iron-sulfur proteins the sulfur (soft) coordinates to both Fe^{2+} (borderline) and Fe^{3+} (hard).

The hard-soft classification scheme is limited to three categories each for acids and bases and does not pretend to be quantitative. It oversimplifies a complicated pattern and as a result its correlations are beset by compromises and inconsistencies. By providing users with a facile vocabulary, it unwittingly misleads them into making frequently incomplete and often incorrect predictions. Biological systems often form complexes that appear ignorant of the scheme. Thus bioinorganic and biochemists would fare better without it. If the oversimplified concept possesses a use, it is as a memory aid or framework on which to peg the correlations that fit, as well as to hang those that do not.

3.6.   pH-Dependent Stabilities

The extent of metal ion binding to ligands in a solution may not depend only upon the stability constant according to the following equation:

$$M + L \rightleftharpoons ML \qquad K_s = [ML] / [M][L]$$

If at a given pH some ligands occur in a protonated form, a competition develops between metal ion and proton for a basic ligand site. The concentration of basic ligand available to the metal ion is reduced by the occurrence of protonated species. The fraction of ligand in the basic form available for coordination is given by $\alpha = K_a / [(H^+) + K_a]$, where the acidity constant K_a is defined in the usual way in terms of the activity of H^+:

$$HL \rightleftharpoons H^+ + L \qquad K_a = (H^+)[L] / [HL]$$

The conditional, pH-dependent stability constant is given by $K_c = \alpha K_s$, or

$$\log K_c = \log K_s + \log(K_a / [(H^+) + K_a]) \qquad (1)$$

For pH $\gg$ pK_a unbound ligand is predominantly in its basic form and $\log K_c = \log K_s$. Neutral solutions of a carboxylate with $pK_a \sim 5$ provide an example. For pH $\ll$ pK_a unbound ligand is predominantly protonated and Eq. (1) reduces to $\log K_c = \log K_s - pK_a + pH$. An example appears in neutral solutions of an aliphatic ammonium ion with $pK_a \sim 10$. When the pH is within two log units of pK_a, the complete Eq. (1) should be employed.

In Table 4 conditional stability constants at pH 7.0 (represented as K_7) are compared with the conventional stability constant (K_s) for three metal ions and four unidentate ligands. The four ligands represent carboxylate, imidazole, sulfhydryl, and aliphatic amine donors in biological systems. For acetate, pH = 7 $\gg$ pK_a, $\alpha = 1$, virtually all ligand is in its basic, carboxylate form, and the conventional and conditional stability constants are equal. For imidazole 44% of the ligand is in the basic form. For glutathione and ammonia only 1.2 and 0.47%, respectively, of the ligands are in the basic form. Glutathione is the tripeptide γ-L-glutamyl-L-cysteinylglycine and it serves as a superb model for the sulfhydryl group in proteins [18,19].

The results in Table 4 show that for all three metal ions the ligand stabilities from the conventional stability constants follow the increasing sequence acetate < ammonia < imidazole < glutathione.

TABLE 4

Conventional and Conditional (pH 7.0) Stability Constants[a]

Ligand	pK_a	Ni^{2+}		Zn^{2+}		Cd^{2+}	
		$\log K_S$	$\log K_7$	$\log K_S$	$\log K_7$	$\log K_S$	$\log K_7$
Acetate (O$^-$)	4.7	0.8	0.8	0.7	0.7	1.2	1.2
Imidazole (=N)	7.11	3.0	2.6	2.52	2.2	2.80	2.4
Glutathione (RS$^-$)	8.92	4.0	2.1	5.0	3.1	6.16	4.2
Ammonia (NH$_3$)	9.33	2.80	0.5	2.37	0.0	2.65	0.3

[a]Conventional stability constants from Ref. 10 for acetate, Ref. 11 for imidazole and ammonia, and Refs. 18 and 19 for glutathione.

This sequence is not the stability order in neutral solutions because the sequence applies only when all ligands are predominantly in their basic form, which is true only for acetate. For the conditional stability constants at pH 7.0, Table 4 shows that for Ni^{2+} the increasing stability order is ammonia < acetate < glutathione < imidazole, while for Zn^{2+} and Cd^{2+} the order is ammonia < acetate < imidazole < glutathione. For all three metal ions in neutral solutions the most basic ammonia has become a weaker ligand than the least basic acetate. The stability order from the conditional constants suggests that in neutral solutions Ni^{2+} is apt to be found at imidazole sites and Zn^{2+} and Cd^{2+} at sulfhydryl sites. Of course chelation by adjacent donors may increase stabilities and alter preferred binding sites. Nonetheless, the examples in Table 4 serve as models for the analogous groups in proteins, and so on, and demonstrate that binding strengths and stability orders are pH dependent. Results of a similar treatment over a greater number of metal ions but without the sulfhydryl group have been presented graphically [20].

Ignored in this section and Table 4 is the formation of metal ion hydroxo complexes and precipitates (see Sec. 3.4). Formulation of conditional stability constants that incorporate hydroxo complex formation is outside the scope of this section.

3.7. Preferred Metal Ion Binding Sites

By incorporating the concept behind the conditional stability constants in the previous sections, we can offer the following generalizations for free metal ion binding sites in neutral aqueous solutions. All alkali, alkaline-earth, and lanthanide metal ions and Al^{3+}, Cr^{3+}, and Fe^{3+} bind at O donor sites. Dipositive metal ions of the first transition row from Mn^{2+} to Cu^{2+} favor sites with increasing numbers of N donors. Zn^{2+} is transitional between N and S donors, depending upon the degree of chelation. Finally, Cd^{2+}, CH_3Hg^+, and Pb^{2+} favor binding at sites with S

donors. These conclusions are helpful in assessing likely binding
sites for metal ions in biological systems, including adventitious
metal ions that may give rise to toxicity. The generalizations do
not intend to and cannot incorporate special cases created by biology, such as Mg^{2+} binding to four nitrogen donors in chlorophyll,
iron in hemes, and iron binding to sulfur donors in iron-sulfur
proteins. On the other hand, these special cases are not examples
of adventitious metal ions and do not lead to toxicity.

3.8. Ligand Exchange Rates

An important but often overlooked feature of metal ions is the rate
of ligand exchange in and out of the metal ion coordination sphere.
The characteristic rate constant for substitution of inner-sphere
water has been determined for many metal ions [21]. Decreasing
water exchange rate constants follow the order $Pb^{2+} \sim Hg^{2+} \sim Cu^{2+} \sim$
alkali metal ions $> Cr^{2+} \sim Cd^{2+} \sim Ca^{2+} >$ lanthanides $\sim Mn^{2+} \sim Zn^{2+} >$
$Fe^{2+} \sim Co^{2+} \sim Mg^{2+} \gg Ni^{2+} > Be^{2+} > Fe^{3+} > Al^{3+} \gg Co^{3+} \ggggg Cr^{3+}$.
Each inequality sign indicates a 10-fold reduction in rate from the
25°C rate constants near 10^9 sec^{-1} at the beginning of the series
and decreasing through 15 powers of 10 to about 10^{-6} sec^{-1} for Cr^{3+}.
Though these specific rate constants represent water exchange in the
aquo metal ions, they also reflect the relative rates of exchange of
other unidentate ligands. Chelated ligands exchange more slowly.

The above series of relative exchange rates shows that some
of the strongest bonding metal ions also undergo rapid ligand
exchange. Examples are Pb^{2+}, Hg^{2+}, Cu^{2+}, and Cd^{2+}. As described
in Sec. 4.19, rapid exchange is an important feature of mercury's
toxicology. Ni^{2+} is not distinguished by stabilities or size
(Table 2) from several metal ions such as Zn^{2+}. The contrast
between the presence of Zn^{2+} in numerous mammalian enzymes and that
of Ni^{2+} in only a few plant enzymes has puzzled many investigators.
The exchange rate series in the previous paragraph provides a reason
for the contrast, as Zn^{2+} exchanges ligands 10^3 times faster than

Ni^{2+}, an important difference for a metal ion bound at enzyme-active
sites. Even slower exchange renders Al^{3+} useless.

4. SURVEY OF METAL IONS

4.1. Introduction

This section applies the general principles already developed to the
toxicity of individual metal ions. For the most part, the order
follows the group in the periodic table. Since essentiality and
toxicity are unrelated, the essential metal ions are not discussed
separately. The emphasis is on toxicity in humans. For many metal
ions *acute* toxicity arising from sudden exposure to a large dose
gives rise to different effects and symptoms than in *chronic* toxicity,
which may be due to a low dose exposure over a long time period. The
symptomatic descriptions of acute and chronic toxicity rely heavily
on two comprehensive compendia [3,5], the contents of which cannot
possibly be covered in a few pages. Reference 5 is an authoritative
guide and should be consulted for further information on individual
metal ions. For a discussion of toxic metals in the environment
see Ref. 22.

The most serious toxic effects of many metal ions are caused
by inhalation of dusts, usually in an industrial setting. Especially
bad are particles 0.1-1 μm in diameter, which are efficiently adsorbed
in the lungs. The lungs are 10 times more effective in absorbing
metal ions into body fluids than are the intestines. Thus the greater
danger from radioactive plutonium-239, an emitter of damaging α particles with a half-life of 24,400 years, results not from its ingestion, but from adsorption of its powders onto lung tissue. Volatile
metal compounds such as the carbonyls and also alkyl compounds of
mercury, tin, and lead are readily absorbed via the lungs and may
give rise to acute toxicity. Thus inhalation of metal ions should
be avoided. However, since these inhalation episodes are relatively
local events, they do not receive as much attention in this chapter

as does the more prevalent and less evident toxicity by ingestion
of metal ions. Toxicity by inhalation is well covered in the
compendia [3,5].

4.2. Alkali Metal Ions

None of the alkali metal ions are especially toxic. Homeostasis
maintains both essential (Table 1) Na^+ and K^+ concentrations within
normal physiological limits. The nutrition of these two elements
has been reviewed [23]. In addition to their specific effects,
metal ions play two general roles in living systems. They contribute to the osmotic balance on the two sides of a membrane. They
also provide the positive counterions for HPO_4^{2-}, HCO_3^-, and organic
molecules, many of which are anionic. The principal extracellular
and intracellular counterions are Na^+ and K^+, respectively. The
other nonessential alkali metal ions may compete with Na^+ and K^+ in
physiological processes, Li^+ being the most toxic. Humans contain
about 0.3 g of Rb^+ associated with K^+ in intracellular fluids.
Little Cs^+ appears in the body, but significant amounts of ^{137}Cs
($t_{1/2}$ = 30 years) have appeared in radioactive fallout. By far the
greatest load of radioactivity to the gonads from normal internal
sources is 20 mrem/year from naturally occurring ^{40}K, unavoidably
and necessarily present in intracellular fluids.

4.3. Lithium

For over 30 years Li^+ has been used in the treatment of manic-
depressive psychoses, with 1 in 2000 people in the United Kingdom
receiving the drug today [24]. The Li_2CO_3 is ingested to attain a
plasma Li^+ concentration of up to 1 mM, markedly reducing mood
changes in many patients. What must be one of the highest levels
of any drug, unfortunately, may produce toxic effects in the forms
of reduced kidney function and central nervous system disorders.

The mode of Li^+ action is still uncertain; it may alter intercellular communication. Addition of Li^+ affects many enzymes, including several involved in glycolysis [24].

Many biochemists believe Li^+ substitutes for Na^+ or K^+, but these metal ions possess about three and six times, respectively, the volume of Li^+, so that any substitutions in proteins result in appreciable restructuring of metal ion cavities. On the other hand, Li^+ is insignificantly larger than Mg^{2+} (Table 2). It usually forms stronger complexes than Na^+ and K^+ but considerably weaker complexes than Mg^{2+}. Under the drug conditions cited above, Li^+ and Mg^{2+} are present in comparable concentrations and Li^+ binds at sites not occupied by Mg^{2+}. If Mg^{2+} sites are saturated, Li^+ substitutes for Na^+ and K^+. All the alkali metal ions undergo exchange reactions more than 10^3 times faster than Mg^{2+} (Sec. 3.8); this factor may explain the alteration in the activity of Mg^{2+} enzymes that contain Li^+.

4.4. Magnesium

Essential magnesium as Mg^{2+} is crucial for both plant and animal life. In plants Mg^{2+} is chelated by four pyrrole-type nitrogens in chlorophyll, a rare instance of Mg^{2+} coordination by nitrogen. In animals Mg^{2+} is an essential enzyme cofactor in virtually every reaction involving adenosine triphosphate (ATP). It also provides a counterion for stabilization of the DNA double helix with its negative phosphate charge load on each residue. The presence of Mg^{2+} increases the fidelity of proper base pairing. When coordinating to nucleoside phosphates such as ATP, Mg^{2+} binds only at the phosphate [25,26]. Mg^{2+} is required for neuromuscular action and muscle contraction. A highly efficient homeostasis maintains plasma Mg^{2+} levels at 0.9 mM in healthy individuals (Table 1).

Mg^{2+} deficiency is more common than recognized, with alcoholism the probable leading cause. Since severe Mg^{2+} deficiency is rare, there is a lack of agreement on its symptoms, apparently

diverse. Such symptoms concentrate on delirium and neuromuscular
manifestations, including tingling, twitching, numbness, and tremors.
Low Mg^{2+} levels may also produce hypocalcemia, in which metabolically
labile mineral is not mobilized from bone. The Mg^{2+} and Ca^{2+} levels
are controlled by a parathyroid hormone through a negative feedback
mechanism.

Magnesium is only slightly toxic. Intake of excessive amounts
of Mg^{2+} salts leads to nausea. Patients with renal malfunctions who
ingest quantities of Mg^{2+} containing antacids may exhibit longer-
term toxic symptoms. These may include a range of central nervous,
respiratory, and cardiovascular system problems.

4.5. Calcium

The two alkali metal ions Na^+ and K^+ and the two alkaline-earth
metal ions Mg^{2+} and Ca^{2+} together make up more than 99% of the metal
ions in the body. Calcium in the form of the ion, Ca^{2+}, is the most
abundant metal ion in the body (Table 1). More than 99% occurs in
the structure of bones and tooth enamel with the composition of
hydroxyapatite, $Ca_5(PO_4)_3(OH)$. In solution Ca^{2+} plays a crucial
role in many processes, including muscle contraction, blood clotting,
neurotransmitter release, microtubule formation, intercellular communication, hormonal responses, exocytosis, fertilization, mineralization, and cell fusion, adhesion, and growth. Many of these Ca^{2+}-
related activities involve interactions with proteins, which Ca^{2+}
may stabilize, activate, and modulate. All known protein binding
sites for Ca^{2+} consist only of oxygen donor atoms. Many aspects of
the biological roles of Ca^{2+} have been reviewed in a recent volume
in this series [27]. A contrast has been drawn between the chemis-
tries of Ca^{2+} and Mg^{2+} applicable to biological systems [28].

The concentration gradient of Ca^{2+} between extracellular and
intracellular fluids far exceeds the gradients of the other three
biologically significant alkali and alkaline-earth metal ions, Na^+,
K^+, and Mg^{2+}. The free Ca^{2+} concentration in extracellular fluids

is about 1.3 mM, while in many intracellular fluids it is an astonishingly low 0.1 µM or less for a 20,000-fold concentration gradient. In response to a stimulus, the low intracellular concentration may increase 10-fold, effecting conformational changes in proteins that possess Ca^{2+} dissociation constants in the micromolar range. The conformational sensitivity of some intracellular proteins to Ca^{2+} concentration changes at the micromolar level leads to the concept of Ca^{2+} as an intracellular, or second, messenger.

The recommended daily intake of 800 mg of Ca^{2+} may be obtained from somewhat less than a liter of milk, the only abundant Ca^{2+} dietary source. Calcium deficiency results in stunted growth, poor teeth, and other defects, some of them less obvious. One of these is the greater absorption of unwanted and toxic metal ions in a Ca^{2+}-deficient system. A homeostatic mechanism that operates by adjusting absorption from the intestine controls the Ca^{2+} level in humans. Calcium is considered nontoxic. Deposit of bone mineral in soft tissues is caused not by excess Ca^{2+}, but by an oversupply of vitamin D. However, high dietary Ca^{2+} levels may inhibit the intestinal absorption of other essential minerals.

4.6. Barium and Strontium

Ba^{2+} is poisonous because of its antagonism to K^+, and not Ca^{2+}. This pairing provides an example of the greater importance of the similar ionic radii of Ba^{2+} and K^+ (see Sec. 3.1 and Table 2) than the identical charge but different radii for the two alkaline-earth metal ions Ba^{2+} and Ca^{2+}. Other examples of Ba^{2+} substitution for K^+ have been described [29,30]. The ion Ba^{2+} is a muscle poison, and treatment consists of intravenous infusion of K^+. While Ba^{2+} is still in the gut, administration of soluble SO_4^{2-} salts yields insoluble $BaSO_4$, which is not absorbed. Thus $BaSO_4$ is employed as x-ray contrast material for gastrointestinal examinations.

Humans contain about 0.3 g of Sr^{2+} in bone. This amount poses no danger unless it is extensively contaminated during years of bone development with ^{90}Sr ($t_{1/2}$ = 28 years) from radioactive fallout.

4.7. Beryllium

As indicated in Sec. 3.4, Be^{2+} forms well below neutrality an insoluble hydroxide, $Be(OH)_2$, that hinders intestinal absorption.
Contact with Be^{2+}-containing salts yields lesions on the skin.
Inhalation of beryllium-containing dusts produces chronic pulmonary
granulomatosis (berylliosis) or nodules in the lung. The condition
develops slowly and is often fatal. Workers on fluorescent lamps,
where the oxide was used as a phosphor, became victims of this
disease. (This application has been discontinued.) Beryllium is
lethal at 1 ppm of body weight.
      Be^{2+} circulates in the body as colloidal phosphate and eventu-
ally becomes incorporated permanently into bone. Formation of
hydroxide and phosphate complexes is consistent with the principles
developed above for the small divalent ion of high charge density.
Be^{2+} inhibits many phosphatase enzymes; it is the strongest known
inhibitor of alkaline phosphatase. It inhibits Mg^{2+}- and K^{+}-
activated enzymes, as well as DNA replication. Chelation therapy
has proven ineffective in removing Be^{2+} from humans with chronic
poisoning. A dangerous substance with latent toxicity, beryllium
should be handled cautiously or not at all.

4.8. Lanthanides

The lanthanides comprise 15 elements from lanthanum at atomic number
57 to lutetium at atomic number 71. All occur only as 3+ ions in
biological systems. As Table 2 indicates for Gd^{3+}, the middle member
of the series (atomic number 64), the ionic radii closely resemble
those of Ca^{2+}. As the likeness in size is more important than the
difference in charge (see Sec. 3.1), lanthanides substitute for Ca^{2+}
in many biological systems. This lanthanide substitution matters
little when the metal ion plays primarily a structural role, but may
be either inhibitory or activating when the metal ion is at the

active site [31,32]. Lanthanide ions have found extensive use as
probes of Ca^{2+} sites in proteins [28,31,32].

   None of the lanthanides are biologically essential. Plants
resist accumulating lanthanides and hence block the transfer of
lanthanides to humans, higher up the food chain. Lanthanides exist
as aqueous 3+ ions up to pH 6, where hydroxo complexes and precipitates begin to form. The phosphates are also insoluble. As a result
lanthanides form insoluble complexes in the gut and are poorly absorbed. None of the lanthanides are considered toxic.

4.9. Aluminum

The most common metal in the earth's crust, aluminum seldom occurs
in living systems, presumably because of its inaccessibility in complex mineral deposits. A typical adult human bears 61 mg of aluminum,
most of it in the lungs from inhalation. The only cation, Al^{3+} in
neutral solutions forms insoluble $Al(OH)_3$ and extensively bridged
hydroxo and oxo species. Formation of these species and insoluble
$AlPO_4$ limits the absorption of ingested Al^{3+}. Following absorption,
the highest concentration occurs in the brain. Impairment of kidney
function reduces the body's ability to clear Al^{3+}. High levels of
Al^{3+} may produce phosphate depletion through formation of $AlPO_4$.

   Only low levels of Al^{3+} have occurred in water and food supplies. At these levels Al^{3+} does not appear especially toxic.
Release of Al^{3+} (and Hg^{2+} and Pb^{2+}) into water supplies by acid
rain may teach us whether higher levels can be a problem. The
released metal ions more than the acidity itself may be toxic to
fish. Limited amounts of Ca^{2+} and Mg^{2+} appear to increase the
likelihood of Al^{3+} toxicity. Harmful effects of Al^{3+} include
constipation and neurotoxicity.

   Increased concentration of Al^{3+} in the brain has been associated with Alzheimer's disease, a malady that brings on dementia
and death to mostly elderly people [33]. Current medical opinion

believes that rather than being the causative factor, the Al^{3+} merely collects in the diseased brain or acts as one contributory factor. Nevertheless, the fact that the elderly use Al^{3+}-based antiperspirants and ingest large quantities of Al^{3+}-containing antacids is a disquieting one. Patients undergoing dialysis with high concentrations of Al^{3+} in the water have developed dialysis dementia.

4.10. Chromium

Lists of essential trace elements traditionally include chromium [34]. The human body contains about 6 mg distributed through many tissues (Table 1). Though the dose requirement has not been established, it must be very small. It has not proved possible to assess chromium status by chemical or biochemical means. The basis for essentiality remains unknown. Though it has been 25 years since Cr^{3+} was first suggested as a component of glucose tolerance factor, the nature of the complex remains unsolved [35]. Some of the structures proposed for the complex are unreasonable.

At pH 7 the favored mononuclear species is $Cr(OH)_2^+$, but highly inert, polynuclear, olated complexes form. Even in hexa-aquo Cr^{3+}, exchange of a water molecule with solvent takes several days (see Sec. 3.8). This inertness appears to limit Cr(III) to a structural role. If it is involved in a rate process, another oxidation state such as Cr(II) must be involved. Sugars offer potentially good ligands for Cr^{3+}. Glucose is a relatively poor metal ion-binding sugar [28], but this limitation may not be important in any Cr^{3+} complexes.

Trivalent Cr(III) is one of the least toxic metal ions; strongly oxidizing hexavalent Cr(VI) is considerably more toxic. At pH < 4, Cr(III) exists as hexa-aquo Cr^{3+}, but as the pH increases, hydroxo and inert polynuclear complexes with bridging oxo groups form. In neutral solutions Cr(VI) exists as CrO_4^{2-}, but the body reduces the strongly oxidizing Cr(VI) to Cr(III). Chemists and

others who handle $CrO_4{}^{2-}$ and acidic $Cr_2O_7{}^{2-}$ solutions should note
their potency for causing skin ulcerations.

4.11. Molybdenum

Molybdenum commonly occurs as Mo(VI), and molybdate, $MoO_4{}^{2-}$, is
absorbed in the gastrointestinal tract (Table 1). Molybdenum occurs
as a cofactor in nitrogenase enzymes in plants. Two Mo, eight Fe,
and two flavin adenine dinucleotide cofactors occur in xanthine
oxidase, the catalyst for the formation of uric acid in animals.
Molybdenum toxicity is related to copper and sulfur status. In
ruminant livestock, feed high in molybdenum and low in copper gives
rise to teart disease, with depressed growth, anemia, and bone disorders. A similar high-Mo, low-Cu combination has been reported to
yield goutlike symptoms in humans. Copper administration usually
benefits animals suffering from molybdenum poisoning. Neither
molybdenum nor the related nonessential wolfram (tungsten), which
inhibits xanthine oxidase activity, is considered especially toxic.

4.12. Manganese

There are several oxidation states available to essential manganese,
but the evidence suggests that it does not function in oxidation-
reduction reactions and only Mn^{2+} is important. Mn^{3+} is not stable
as an aqueous ion at pH > 0 and must be complexed to avoid reduction
to Mn^{2+} in neutral solutions. No strong evidence exists for manganese deficiency in humans. In animals deficiency results in
impaired bone growth, reduced reproductive function, and possible
depressed cholesterol synthesis. Manganese may serve as an enzyme
cofactor. Though many enzymes are activated by Mn^{2+}, the activation is not specific, because other metal ions such as Mg^{2+} are
also effective. The plasma concentration of Mn^{2+} is only 1/1000
that of Mg^{2+} (Table 1).

Manganese is almost nontoxic, especially in the Mn^{2+} state.
Because it is strongly oxidizing, permanganate, MnO_4^-, is toxic.
The most common manganese toxicity results from inhalation of manganese oxides in an industrial setting. Chronic exposure produces
manganism, in which there is serious, irreversible damage to the
central nervous system and brain. Evidently excess manganese
affects enzymatic systems in the brain. There is no universally
effective antidote, and the condition persists upon withdrawal of
the initial cause. Some success has been achieved by treating
patients in the early stages with L-dopa (L-3,4-dihydroxyphenyl-
alanine) [36].

4.13. Iron

A typical human contains about 4 g of iron (Table 1), of which about
70%, or 3 g, appears within the red cells in hemoglobin. Most of the
remainder occurs in iron storage proteins, with a small amount in a
few enzymes. Of the recommended daily allowance of 10-20 mg of iron,
only 10-20% is absorbed, the higher amount occurring in iron-deficient
individuals under fine homeostasis. Absorption is inhibited by the
formation of insoluble hydroxide, phosphate, and fatty acid complexes
and aided by soluble sugar and ascorbate chelates [37]. Almost all
of the 25 mg of iron released daily by hemoglobin catabolism is efficiently recycled in the liver, so that the half-life of iron in the
body may exceed 10 years. Therefore absorption of less than 1 mg
daily is sufficient for all but menstruating women, who lose about
20 mg of iron each period. The world's most common human deficiency
is that of iron, affecting 10% of premenopausal women in industrial
countries and rising to 100% in some groups. Iron deficiency leads
to anemia.

Iron is absorbed in the Fe(II) form and oxidized to Fe(III)
in the blood. Since Fe^{3+} forms highly insoluble precipitates even
in acidic aqueous solutions (see Sec. 3.4), a protein transferrin

carries Fe^{3+} in the blood. When the Fe^{3+}-carrying capacity of
transferrin is exceeded, $Fe(OH)_3$ precipitates in the blood.

Iron toxicity occurs in specific groups: among thousands of
children, an estimated 10 of whom die each year in the United States
from swallowing mineral supplement pills high in $FeSO_4$ intended for
their mothers; among the Bantus, who cook and brew in iron pots; and
among some alcoholics who have suffered extreme liver damage. Iron
toxicity gives rise to lesions in the gastrointestinal tract, shock,
and liver damage.

4.14. Cobalt

Cobalt occurs as an essential component of vitamin B_{12} chelated in
a complex corrin macrocycle with four linked pyrrole rings [11].
The body needs only a miniscule 3 μg of vitamin B_{12} daily (Table 1).
A deficiency results in anemia and stunted growth. There are several forms of vitamin B_{12} that serve as enzyme cofactors in methyl
transfer and other reactions where the cobalt undergoes a change
in oxidation state.

When not bound in the vitamin B_{12} corrinoid ring, cobalt
exists in biological systems as Co^{2+}, an ion able to accommodate
four, five, or six donor atoms in the variable coordination poly-
hedron. Zn^{2+} possesses an identical capability, and the two ions
exhibit similar effective ionic radii in all coordination numbers
(Table 2) and display comparable stability constant values with
many ligands (Table 3). Co^{2+} replaces Zn^{2+} in several Zn^{2+} metalloenzymes, often giving an active enzyme. Because it contains unpaired
d electrons, several spectroscopic methods make Co^{2+} a useful probe
of the spectroscopically silent Zn^{2+} in Zn^{2+} metalloproteins [38].
metalloproteins [38].

Excess Co^{2+} stimulates red bone marrow to increase the number
of red blood cells; such excess also decreases the ability of the
thyroid to accumulate iodine, so that goiter may occur as a possible
side effect of Co^{2+} administration for anemia.

In a curious and unfortunate episode with a nearly 50% mortality rate, Co^{2+} induced cardiotoxicity among some, but not all,
heavy beer drinkers (more than 3 liters/day): In several countries
about 1 ppm of a Co^{2+} salt had been added to beer to stabilize the
foam against the effect of residual detergent. Even though the
amounts ingested corresponded to less than those administered to
anemic patients, evidently ethanol sensitized the body to Co^{2+}
intoxication and the SO_2 contained in bottled beer destroyed
thiamine, whose deficiency aggravated Co^{2+}-induced cardiotoxicity
[39].

4.15. Nickel

In biological systems nickel occurs in solution almost exclusively
as Ni(II). Though an accessible oxidation state under some conditions, Ni(III) is unlikely to be important in higher organisms [40].
The human body contains about 10 mg of Ni^{2+}, and plasma levels hold
within a narrow range, suggesting a homeostatic mechanism and possible essentiality (Table 1). At the minimum, low levels of Ni^{2+}
appear stimulating in animals. It serves as a cofactor in the plant
enzyme urease. With other metal ions Ni^{2+} activates some animal
enzymes, but it has not been established as essential in humans.

Nickel as Ni^{2+} is another metal ion that is relatively nontoxic; however, industrial fumes, especially of nickel carbonyl,
$Ni(CO)_4$, where nickel is formally in the zero oxidation state, are
easily absorbed by the lungs and are highly toxic. Ingested Ni^{2+}
causes acute gastrointestinal discomfort. Chronic Ni^{2+} toxicity
leads to degeneration of heart and other tissues. The toxic action
of Ni^{2+} is unknown; it blocks enzymes and combines with nucleic
acids. Sweating can leach Ni^{2+} from metal-plated jewelry and cause
dermatitis. Ni^{2+} complexes with the skin protein keratin. Perhaps
this is the most common skin-sensitizing reaction among women.

4.16. Copper

Copper concentration is regulated by homeostasis, and the optimal
concentration range across the top of Fig. 1 is broad. Therefore
neither copper deficiency nor toxicity is of general frequent concern. Copper is an essential (Table 1) cofactor in several enzymes,
catalyzing various oxidation-reduction reactions. Its deficiency
results in anemia, poor bones and connective tissue, and loss of
hair pigmentation. It is possible that a high Zn^{2+} intake such as
from pills may induce a copper deficiency.

Both valence states of copper, Cu(I) and Cu(II), bind to sulfhydryl groups that occur in glutathione and proteins. The Cu(II)
oxidizes unhindered sulfhydryl groups to disulfides, itself being
reduced to Cu(I). Organisms need to sequester Cu(II) before sulfhydryl oxidation occurs. About 95% of plasma copper is carried by
the protein ceruloplasmin. Even though it carries one sulfhydryl
group, the primary Cu^{2+}-binding site in neutral solutions of plasma
albumin is the amino terminus, where the amino nitrogen, two depro-
tonated peptide nitrogens, and an imidazole side chain nitrogen of
the third amino acid residue chelate Cu^{2+} tightly in a planar ring
system [40]. Hexa-aquo Cu^{2+} becomes more tetragonal (planar) as
the number of nitrogen donors increases.

An acute excess of ingested copper irritates the stomach nerve
endings and leads to nausea. A chronic excess results in reduced
growth, hemolysis and low hemoglobin, and tissue damage to liver,
kidney, and brain. Ceruloplasmin is lacking in most patients with
Wilson's disease, an inborn metabolism error. Wilson's disease
patients exhibit elevated copper levels in the liver along with
liver dysfunction. Copper toxicity may be reduced by ingesting
more MoO_4^{2-} (see molybdate, Sec. 4.11), SO_4^{2-}, and Zn^{2+}.

4.17. Zinc

In humans Zn^{2+} occurs in over 20 metalloenzymes, including several
involved in nucleic acid metabolism. Most of the Zn^{2+} in blood is
found in the red cells as an essential (Table 1) cofactor in the
enzyme carbonic anhydrase. There is only the one oxidation state
in solution. The role of Zn^{2+} in enzymes involves either its direct
binding to and polarization of a substrate or an indirect interaction
through bound water and hydroxide as general acid-base catalysts and
nucleophiles [38]. Most of the body Zn^{2+} appears in muscle, but the
highest concentration occurs in the prostate gland. Levels of Zn^{2+}
are under homeostatic control.

Zinc deficiencies occur among some alcoholics and are prevalent
in developing countries where diets are high in absorbing fiber or
clay. Zinc deficiency results in skin disorders, growth retardation,
and reduced sexual development and function in young men. Though
there is no known human aphrodisiac, adequate Zn^{2+} appears essential
for male sexual function. Because human spermatogenesis is a lengthy
process, improving any reduced sexual function by increasing Zn^{2+}
levels takes time. There is no evidence that excess Zn^{2+} beyond an
optimal level produces any additional effects. On the contrary, Zn^{2+}
supplements may unbalance metabolism of other minerals and should be
taken only under medical supervision. This advice is especially
sound if the hypothesis of a high Zn^{2+}-to-Cu^{2+} ratio as the main
causative factor in ischemic heart disease (locally obstructed arte-
rial blood flow) proves correct [41]. A Zn^{2+} supplement speeds wound
healing in Zn^{2+}-deficient patients, but is without effect when adequate Zn^{2+} is present. Zinc occurs so plentifully in meat and shellfish that supplements are unnecessary for residents of industrial
countries and may even be harmful to the extent that they reduce the
absorption of copper, iron, and other essential metal ions.

Ingestion of excess Zn^{2+} salts may result in acute gastrointestinal disturbances with nausea. Instances of acute toxicity
have occurred from the ingestion of acidic fruit juices that were
stored in galvanized (zinc plated) steel containers. Chronic

toxicity due to Zn^{2+} in humans is virtually unknown, but there may be subtle, unrecognized effects. For example, by competition, excess Zn^{2+} may cause a Cu^{2+} deficiency if that ion is present in only minimal amounts. Similarly, excess Zn^{2+} appears to hinder bone development in animals when calcium and phosphorus are present in marginal amounts. Overall, Zn^{2+} is not especially hazardous, and its greatest danger is probably from associated toxic Cd^{2+} as an impurity.

4.18. Cadmium

Relatively rare, Cd^{2+} occurs in minerals and soils in association with zinc to about 0.1%. Like Zn^{2+}, the metal occurs only as a 2+ ion, Cd^{2+}. Cd^{2+} is larger than Zn^{2+} and more nearly the size of Ca^{2+} (Table 2), leading to its use as a Ca^{2+} probe [28]. However, in its binding strength to ligands Cd^{2+} is much more like Zn^{2+} (Table 3), and it is in competition with Zn^{2+} that the greater number of its toxic effects apparently arise. In contrast to Ca^{2+}, both metal ions bind strongly to nitrogen and sulfur donor ligands (see Sec. 3.2 and 3.3). Excess Cd^{2+} disturbs mineral metabolism, interferes with Zn^{2+} and other enzymes [42], and may redistribute Zn^{2+} in the body. The exact mechanisms of Cd^{2+} toxicity (certainly manifold) are unknown.

In strong contrast to CH_3Hg^+, Cd^{2+} does not easily cross the placental barrier, and the newly born are virtually Cd^{2+} free. For most individuals Cd^{2+} is slowly accumulated from foods. The body only slowly releases absorbed Cd^{2+}, which has a half-life of more than 10 years. As a consequence, the Cd^{2+} content of the kidney increases throughout life from zero at birth to about 20 mg in a middle-aged nonsmoker and more than twice as much in an adult smoker. Much of the Cd^{2+} is bound by metallothionein, a small protein with sulfhydryl-binding sites, whose production is stimulated by Cd^{2+}.

Acute Cd^{2+} poisoning signaled by vomiting, abdominal cramps, and headache, may arise from drinking water or other, especially acidic, liquids that have contacted cadmium-containing solders in water pipes and vending machines or cadmium-glazed pottery. Ingested Cd^{2+} is transported to other organs by the blood, where it is bound by glutathione and hemoglobin in the red cells [43]. The blood of smokers contains about seven times as much Cd^{2+} as that of nonsmokers. Chronic Cd^{2+} poisoning damages the liver and kidneys, leading to severe kidney malfunction. There is no specific therapy for Cd^{2+} poisoning; chelating agents may harmfully redistribute Cd^{2+} to the kidney. Plentiful intakes of Zn^{2+}, Ca^{2+}, phosphate, vitamin D, and protein may lessen the impact of Cd^{2+} toxicity.

An especially serious form of Cd^{2+} poisoning has been described in Japan as itai-itai (Japanese for ouch-ouch) disease. The name derives from pain in the back and legs due to osteomalacia or bone decalcification (usually in older women) leading to easy fracturing (72 fractures in one victim). There is also severe kidney damage with proteinuria (protein in the urine), persisting even after exposure ceases. This disease leads to early death. In Japan it was produced by Cd^{2+}-laden water, either directly by drinking or by eating crops, especially rice, irrigated by the water over a period of many years. Though there was earlier, visible damage to the rice crop, it took 17 years from the first medical studies on humans to the final fixing of legal responsibility on a zinc mining company that discarded the unwanted cadmium by-product into a river used for drinking and irrigation [44,45].

4.19. Mercury

Mercury is toxic in all of its forms. The global release of mercury by degassing from the earth's crust and oceans exceeds that produced by humans by a factor of at least five, but industrial discharges are more local and concentrated. The average person carries 13 mg of mercury in the body, none of which serves any useful purpose. Various mercuric salts have been used as therapeutic agents, for example, mercuric benzoate to treat syphilis and gonorrhea. Use of

mercury reagents as insecticides and fungicides has led to many
small-scale and several disastrous large-scale poisonings involving
thousands of individuals. Mercury poisoning is a worldwide problem.

4.19.1. Chemistry

Mercury occurs in three common forms and one uncommon form as
mercurous ion that disproportionates to two of the common forms,
elemental mercury and mercuric ion:

$$Hg_2^{2+} \rightleftarrows Hg^0 + Hg^{2+}$$

The equilibrium constant value for the reaction

$$K = [Hg^{2+}] / [Hg_2^{2+}] \simeq 1/170$$

indicates that the favored direction of the reaction is actually
from right to left. The reaction is driven to the right by strong
complexation of Hg^{2+} with any of a wide number of ligands. The
third common form of mercury is as organic mercurials such as methyl
mercury, CH_3Hg^+.

Mercury is one of the few elements that is liquid at room
temperature. Though its boiling point is 357°C, it is more volatile
and hence more dangerous than generally appreciated. At saturation
at 25°C a cubic meter of air holds 20 mg of Hg. Spilled mercury is
a health hazard in rooms with poor ventilation. Elemental mercury
is almost insoluble in water, dissolving to the extent of 0.28 μM
at 25° [46], 56 μg/liter, or 56 ppb.

Both mercuric ion, Hg^{2+}, and methyl mercury, CH_3Hg^+, prefer
linear, 2-coordination. They form stronger complexes than most
metal ions with a single donor atom, especially N and S (see Sec.
3.3). Alone among the metal ions considered in this chapter, Hg^{2+}
in basic solutions substitutes for a hydrogen on amines (as opposed
to ammonium ions) as in mercuric amidochloride, $ClHgNH_2$. At high
pH values excess CH_3Hg^+ substitutes for a hydrogen on the $-NH_2$
group at C4 in cytidine and C6 in adenosine [25]. Hg^{2+} may add
third and fourth donor atoms to give a distorted tetrahedron. The
last two ligands are bound much more weakly than the first two and

exhibit longer Hg^{2+}-donor atom bond lengths. The ability of linear, 2-coordinated Hg^{2+} and CH_3Hg^+ to associate weakly with additional donor atoms becomes important in accounting for the rapid metal ion exchange among donor atoms, discussed below.

Because they prefer linear, 2-coordination, there is not much of a chelate effect with Hg^{2+} and CH_3Hg^+. In contrast to most metal ions, the stability constant logarithm for Hg^{2+} and CH_3Hg^+ binding to 1,2-diaminoethane is less than the sum of the values for binding to two ammonia molecules, that to glycine is less than the sum of ammonia and acetate, and that to histidine is less than the sum of ammonia and imidazole. However, owing to the great strength of binding to a single nitrogen donor atom, the stability constants for the weakly chelated Hg^{2+} and CH_3Hg^+ with diaminoethane, glycine, and histidine still exceed those for most metal ions (see Secs. 3.2 and 3.3).

In pure water HgO begins to precipitate at pH > 2 and CH_3HgOH forms at pH > 5. Chloride ion is bound with high stability. The first two Cl^- to give linear $HgCl_2$ are bound 10^6 times more strongly than the last two Cl^- to give tetrahedral $HgCl_4^{2-}$ [14]. Because of the strong chloride binding, in seawater, with its high 0.56 M Cl^- background, the predominant species become $HgCl_4^{2-}$ and CH_3HgCl.

In almost any biological system an excess of sulfhydryl donors is expected to bind all Hg^{2+} and CH_3Hg^+. In fact the name mercaptan derives from the strong Hg^{2+} binding to thiols. In red blood cells Hg^{2+} links glutathione and hemoglobin sulfhydryl groups in a mixed complex [47]. Only a fraction of body Hg^{2+} remains in the blood. Though the molecular basis of Hg^{2+} toxicity by interaction at sulf-hydryl groups is generally understood, the specific proteins under-going consequential damage remain uncertain. Many proteins contain reactive sulfhydryl groups. Hg^{2+} also forms inert bonds at C5 in pyrimidines [48]. It is possible that the long retention times of mercury in tissues may result from the formation of inert aromatic carbon-mercury bonds involving nucleic bases or amino acid side chains.

An interesting and significant feature of Hg^{2+} and CH_3Hg^+ complexes is rapid exchange of ligands in and out of the coordination sphere. Substitution of inner-sphere water in Hg^{2+} occurs with a rate constant of $10^9 \ sec^{-1}$, as fast as any metal ion (see Sec. 3.8). For CH_3Hg^+ binding to the sulfhydryl group of glutathione the stability constant logarithm K = 15.9 [49]. For such strong binding the half-life for exchange by a dissociation mechanism is about 10 days, yet the average lifetime for the CH_3HgSR complex in the presence of excess glutathione is less than 0.01 sec [50]. This comparison shows that exchange does not occur by a dissociation mechanism. The clue to the exchange mechanism in Hg^{2+} and CH_3Hg^+ compounds is provided by weak coordination of third and fourth ligands (mentioned near the beginning of this section). Excess ligand participates in a nucleophilic attack at an uncoordinated site on the metal ion, with rearrangement in the coordination geometry and release of a formerly bound ligand. This addition-elimination mechanism accounts for the very rapid exchange in Hg^{2+} and CH_3Hg^+ complexes.

The rapid exchange of Hg^{2+} and CH_3Hg^+ among excess donor ligands such as sulfhydryl groups is of enormous importance in toxicology. It accounts for the rapid partitioning of the mercury among sulfhydryl groups in tissues. In the blood CH_3Hg^+ partitions in the same proportion as the populations of sulfhydryl groups, about 10% in the plasma and 90% in the erythrocytes, which contain both hemoglobin and glutathione sulfhydryl groups. To be avoided as an antidote for mercury poisoning is simple administration of BAL (2,3-dimercaptopropanol), as it facilitates the distribution of mercury throughout the body. Instead hemodialysis with infusion of chelating agents such as cysteine or N-acetylpenicillamine is required.

4.19.2. Toxicity

Inhaled mercury vapor is efficiently absorbed and accumulates in the brain, kidneys, and testicles. It also crosses the placental barrier.

Acute poisoning causes lung damage. In tissues elemental mercury is converted to mercuric ion, which combines with sulfhydryl-containing molecules, including proteins. Chronic poisoning gives rise to permanent central nervous system damage, resulting in fatigue and at higher levels to a characteristic mercurial tremor in which a fine trembling is interrupted every few minutes by coarse shaking.

Ingestion of only 1 g of a mercuric salt is lethal. Mercuric salts accumulate in the kidney; in contrast to elemental mercury, they do not readily cross the blood-brain or placental barriers. Acute poisoning by ingestion results in precipitation of proteins from the mucous membranes of the gastrointestinal tract, causing pain, vomiting, and diarrhea. If the patient survives this damage, the kidney becomes the critical organ. There is some red cell hemolysis. Chronic poisoning gives rise to central nervous system damage. The Mad Hatter in *Alice in Wonderland* by Lewis Carroll represents an occupational disease suffered from poisoning by $Hg(NO_3)_2$ used in the treatment of furs.

Organic mercuricals such as methyl mercury chloride, CH_3HgCl, are highly toxic owing to their high volatility. Microorganisms in polluted water freely convert various forms of inorganic mercury to methyl mercury CH_3Hg^+. Most of the mercury in fish occurs as CH_3Hg^+, which is retained for years or until the fish is eaten. High levels of CH_3Hg^+ do not appear as toxic to fish as to humans. In humans inhaled or ingested CH_3Hg^+ is efficiently absorbed, appears in the red blood cells, liver, and kidney, and accumulates in the brain, including the fetal brain, where it causes serious, cumulative, irreversible damage to the central nervous system. In the human body the retention half-life ranges from months to years. The effects are insidious and the appearance of symptoms often delayed for many years.

Two of the most well-known examples of mercury poisoning were both due primarily to CH_3Hg^+. In 1956 "Minamata disease" was discovered near Minamata bay in southern Japan. In 1959 it was shown that the disease was caused by eating fish contaminated by mercury from CH_3HgCl discharged by a chemical manufacturer into the bay. The mercury load was so great that some fish died, fish-eating birds dropped into the sea, and cats would stagger, "whirl violently, then dash away on a zigzag course or collapse" [51]. Even as early as 1954 this "dancing disease" had markedly reduced the local cat population. No records of mercury contamination levels in the area were taken before 1959. Owing to the Japanese custom of saving dried umbilical cords of infants, it was possible to show human mercury contamination as far back as 1947. Not until 1968 was the dumping brought to a halt.

In humans onset of Minamata disease due to CH_3Hg^+ begins with numbness in the limbs and face, sensory disturbance, and difficulty with hand movements such as in writing. Later there is a lack of coordination, weakness, tremor, uncoordinated walk, mental disturbance, slurred speech, impaired hearing, and constriction of the visual field. Finally, there is general paralysis, limb deformity, especially in the fingers, difficulty in swallowing, convulsions, and death [51]. Tragically also, children born of mildly diseased mothers, who may or may not have exhibited symptoms themselves, suffered from cerebral palsy and idiocy. (Normally cerebral palsy is not associated with significant mental retardation.) Evidently the mother discharges CH_3Hg^+ through the placental barrier into the highly susceptible fetus [51]. Women with more serious disease levels are unable to give birth to live children.

In Iraq in the winter of 1971-1972 ingestion for 2-3 months of bread made from seed wheat treated with a CH_3Hg^+ fungicide resulted in over 6000 hospital admissions and 500 deaths. Symptoms were similar to those of Minamata disease listed above. In the United States registration of CH_3Hg^+ as a seed fungicide was suspended in 1970, but a manufacturer obtained an injunction against the suspension, allowing sales to continue. A suspension by the Environmental Protection Agency was finally made to stick in 1972 [52].

4.20. Thallium

Absorption into the body of the extremely toxic thallium compounds
leads to gastroenteritis, peripheral neuropathy, and often death.
In longer-lasting chronic cases hair loss occurs. Use of Tl_2SO_4
as a rodenticide has been curtailed owing to its high toxicity and
appearance in other wild and domestic animals. The main form in
the body is Tl^+, though the TlCl ion pair is only slightly soluble
and Tl^{3+} also exists. Tl^+ is only slightly larger than K^+ (see
Sec. 3.1) and probably owes much of its toxicity to competition
with K^+ and its ability to permeate cell membranes as K^+. Though K^+
and Tl^+ are similar in size, Tl^+ is almost four times more polarizable than K^+ and forms stronger complexes. It forms an insoluble
riboflavin complex and may also interfere with sulfur metabolism.
In an exception to the lower oxidation state being softer in the
hard-soft classification (see Sec. 3.5), Tl^+ behaves as harder than
Tl^{3+} [15].

4.21. Lead

Lead has been known for almost 5000 years and its toxicity was
recorded by Greek and Arab scholars. There was high lead toxicity
among wealthy Romans, as their wine was stored and their food cooked
in lead vessels. Goya, like other painters, suffered from inhalation
and accidental ingestion of lead in paints. Today elevated lead
levels occur in urban children when they nibble on objects painted
with a lead-based pigment, play in junkyard dirt where lead content
is high from discarded batteries, make spitballs out of magazine
pages that contain 0.4% lead in the colored newsprint, and most of
all breathe air contaminated by the combustion products of tetraethyl lead, $Pb(C_2H_5)_4$, used as an antiknock agent in petrol. Moreover wildfowl hunting along flyways contaminates marshes and lakes
with lead shot that is poisoning plants, birds, and mammals. Lead
oxidizes to Pb^{2+} under the acidic conditions in the stomach.

Generally, the main source of lead intake is in the diet.
Fortunately absorption of ingested lead is low owing to formation
of insoluble lead phosphate, $Pb_3(PO_4)_2$, and basic lead carbonate,
$Pb_3(CO_3)_2(OH)_2$. Absorbed lead accumulates in the bone, where it is
released later in life due to osteoporosis, leading to a possible
delayed toxicity. Today the average adult body contains about 120 mg
of lead, 10 times the amount found in Egyptian mummies. In the absence of anions that cause precipitation, the pH 7 form is Pb^{2+}.
The internationally agreed upon maximum admissible concentration of
lead in drinking water is 50 μg/liter.

Acute lead toxicity produces appetite loss and vomiting.
Chronic toxicity leads to renal malfunction, anemia, and nervous
system disorders, including brain damage in children. The effects
are more serious for a patient deficient in calcium or iron. Lead
interferes with virtually every one of many steps in the biosynthesis
of heme [5]. Other than this damage, it is difficult to specify
critical interactions in lead toxicity.

Pb^{2+} binds strongly to ligands containing sulfhydryl groups,
but not so emphatically as Hg^{2+} (see Sec. 3.3). Some forms of Pb^{2+}
poisoning are treated effectively by administering CaEDTA. Aspects
of lead toxicity appear in other chapters in this volume and in
Refs. 22, 42, and 53.

5. CAVEAT

Though a useful and instructive procedure, singling out properties
for comparison among metal ions as done in Sec. 3 oversimplifies
metal ion interaction modes in complex biological systems. For
example, Sec. 3.1 compares metal ionic radii and points up opportunities for isomorphous replacement of a natural metal ion by
another. But a toxic metal ion is not limited to isomorphous
replacement: It may force a nonisomorphous replacement for a naturally occurring metal ion or coordinate at a crucial site not normally occupied by a metal ion. These actions may significantly

62   MARTIN

alter active sites and conformations. Such "wrecker" roles of nonisomorphous replacements and coordination by potentially toxic metal
ions are difficult to predict specifically but usually appear reasonable when discovered empirically.

In biological systems complexities of the kind referred to
above are compounded by several routes by which metal ions enter an
organism. Metal ions may enter the body by injection, skin absorption, inhalation, and ingestion. This article has ignored entry
through injection. Few metals enter the body through the skin.
Inhalation of dusts is the major entry route in an industrial setting, but volatile metal compounds have a wide distribution. Inhaled metals are often adsorbed in the lungs and by subsequent
absorption distributed to other organs. For most people ingestion
is the likely route for entry by toxic metal ions. Fortunately,
many metal ions are poorly absorbed from the gastrointestinal tract.

An additional compounding feature of metal ion toxicity
already mentioned repeatedly is the competitive interrelationships
among metal ions and cooperative interactions with some anions that
may give rise to variable toxicities for comparable concentrations
of a designated metal ion. Again, such interrelationships are often
difficult to anticipate but when found appear consistent with current
knowledge.

Though we recognize from the start complexities associated
with much metal ion toxicity, the comparisons among metal ions made
in this chapter show that we have begun to understand and may eventually be able to predict the precise molecular bases for toxicities
of a specific metal ion.

REFERENCES

1. "Inorganic Drugs in Deficiency and Disease," *Metal Ions in Biological Systems,* Vol. 14 (H. Sigel, ed.), Marcel Dekker, New York, 1982.
2. E. J. Underwood, *Trace Elements in Human and Animal Nutrition,* 4th ed., Academic Press, New York, 1977.

3. B. Venugopal and T. D. Luckey, *Metal Toxicity in Mammals,* Vols. 1 and 2, Plenum Press, New York, 1977 and 1978.

4. M. Vašak and J. H. R. Kägi, *Metal Ions Biol. Syst., 15,* 213 (1983).

5. L. Friberg, G. F. Nordberg, and V. B. Vouk, *Handbook on the Toxicology of Metals,* Elsevier, Amsterdam, 1979.

6. "Carcinogenicity and Metal Ions," *Metal Ions in Biological Systems,* Vol. 10 (H. Sigel, ed.), Marcel Dekker, New York, 1980.

7. "Metal Complexes as Anticancer Agents," *Metal Ions in Biological Systems,* Vol. 11 (H. Sigel, ed.), Marcel Dekker, New York, 1980.

8. "Micronutrient Interactions: Vitamins, Minerals, and Hazardous Elements," *Ann. N.Y. Acad. Sci., 355* (1980).

9. R. D. Shannon, *Acta Crystallogr. A, 32,* 751 (1976).

10. A. E. Martell and R. M. Smith, *Critical Stability Constants,* Vols. 1-5, Plenum Press, New York, 1974-1982.

11. R. J. Sundberg and R. B. Martin, *Chem. Rev., 74,* 471 (1974).

12. R. B. Martin, *Metal Ions Biol. Syst., 9,* 1 (1979).

13. R. B. Martin, *Proc. Nat. Acad. Sci. U.S.A., 71,* 4346 (1974).

14. C. F. Baes, Jr., and R. E. Mesmer, *The Hydrolysis of Cations,* John Wiley, New York, 1976.

15. R. G. Pearson (ed.), *Hard and Soft Acids and Bases,* Dowden, Hutchinson, and Ross, Stroudsberg, Pennsylvania, 1973.

16. S. Ahrland, J. Chatt, and N. R. Davies, *Q. Rev. Chem. Soc., 12,* 265 (1958).

17. E. Nieboer and W. A. E. McBryde, *Can. J. Chem., 51,* 2512 (1973).

18. R. B. Martin and J. T. Edsall, *J. Am. Chem. Soc., 81,* 4044 (1959).

19. D. L. Rabenstein, R. Guevremont, and C. A. Evans, *Metal Ions Biol. Syst., 9,* 103 (1979).

20. H. Sigel and D. B. McCormick, *Acct. Chem. Res., 3,* 201 (1970).

21. H. Diebler, M. Eigen, G. Ilgenfritz, G. Maass, and R. Winkler, *Pure Appl. Chem., 20,* 93 (1969).

22. F. W. Oehme (ed.), *Toxicity of Heavy Metals in the Environment,* Parts 1 and 2, Marcel Dekker, New York, 1978.

23. M. J. Fregly, *Annu. Rev. Nutr., 1,* 69 (1981).

24. N. J. Birch, *Metal Ions Biol. Syst., 14,* 257 (1982); see Ref. 1.

25. R. B. Martin and Y. H. Mariam, *Metal Ions Biol. Syst., 8,* 57 (1979).

26. Y. H. Mariam and R. B. Martin, *Inorg. Chim. Acta, 35,* 23 (1979).

27. "Calcium and Its Role in Biology," *Metal Ions in Biological Systems,* Vol. 17 (H. Sigel, ed.), Marcel Dekker, New York, 1984.

28. R. B. Martin, *Metal Ions Biol. Syst., 17,* 1 (1984); see Ref. 27.

29. K. Hermsmeyer and N. Sperelakis, *Am. J. Physiol., 219,* 1108 (1970).

30. N. Sperelakis, M. F. Schneider, and E. J. Harris, *J. Gen. Physiol., 50,* 1563 (1967).

31. R. B. Martin, in *Metal Ions in Biology,* Vol. 6 (T. G. Spiro, ed.), John Wiley, New York, 1983, pp. 235-270.

32. R. B. Martin and F. S. Richardson, *Q. Rev. Biophys., 12,* 181 (1979).

33. R. J. Wurtman, *Sci. Am., 252,* No. 1, 62 (1985).

34. K. M. Hambridge, in *Disorders of Mineral Metabolism,* Vol. 1 (F. Bronner and J. W. Coburn, eds.), Academic Press, New York, 1981, p. 271.

35. W. Mertz, *Science, 213,* 1332 (1981).

36. I. Mena, in *Disorders of Mineral Metabolism,* Vol. 1 (F. Bronner and J. W. Coburn, eds.), Academic Press, New York, 1981, p. 233.

37. L. Hallberg, *Annu. Rev. Nutr., 1,* 123 (1981.

38. "Zinc and Its Role in Biology and Nutrition," *Metal Ions Biological Systems,* Vol. 15 (H. Sigel, ed.), Marcel Dekker, New York, 1983

39. Entire October 7 issue, *Can. Med. Assoc. J., 97,* 881 (1967).

40. H. Sigel and R. B. Martin, *Chem. Rev., 82,* 385 (1982).

41. L. M. Klevay, in *Metabolism of Trace Metals in Man,* Vol. 1 (O. M. Rennert and W. Chan, eds.), CRC Press, Boca Raton, Florida, 1984, p. 129.

42. B. L. Vallee and D. D. Ulmer, *Annu. Rev. Biochem., 41,* 91 (1972).

43. D. L. Rabenstein, A. A. Isab, and P. Mohanakrishnan, *Biochim. Biophys. Acta, 762,* 531 (1983).

44. J. Kobayashi, in *Toxicity of Heavy Metals in the Environment,* Part 1 (F. W. Oehme, ed.), Marcel Dekker, New York, 1978, pp. 199-260.

45. L. Friberg, M. Piscator, G. F. Nordberg, and T. Kjellström, *Cadmium in the Environment,* 2nd ed., CRC Press, Cleveland, 1974.

46. J. N. Spencer and A. F. Voigt, *J. Phys. Chem., 72,* 464 (1968).

47. D. L. Rabenstein and A. A. Isab, *Biochim. Biophys. Acta, 721,* 374 (1982).

48. R. M. K. Dale, E. Martin, D. C. Livingston, and D. C. Ward, *Biochemistry, 14,* 2447 (1975).

49. D. L. Rabenstein, *Acct. Chem. Res., 11,* 100 (1978).

50. D. L. Rabenstein, A. A. Isab, and R. S. Reid, *Biochim. Biophys. Acta, 696,* 53 (1982).

51. M. Harada, in *Toxicity of Heavy Metals in the Environment,* Part 1 (F. W. Oehme, ed.), Marcel Dekker, New York, 1978, pp. 261-302.

52. D. R. Cassidy and A. Furr, in *Toxicity of Heavy Metals in the Environment,* Part 1 (F. W. Oehme, ed.), Marcel Dekker, New York, 1978, pp. 303-330.

53. *Lead in the Human Environment,* National Academy of Sciences, Washington, D.C., 1980.

3

The Interrelation Between Essentiality and Toxicity of Metals in the Aquatic Ecosystem

Elie Eichenberger
Federal Institute for Water Resources and Water Pollution Control
Department of Technical Biology
EAWAG's Experiment Station Tüffenwies 36, CH-8064
Zurich, Switzerland

1. GENERAL INTRODUCTION

Metals play a double role in the physiology of microorganisms,
plants, and animals. While some metals are indispensable in normal
development and survival, they all become toxic at relatively low
concentrations. In the aquatic environment these toxic properties
have become a major focus of concern. Metal deficiency, on the
other hand, has been considered a possible factor limiting primary
production. Pollution of the aquatic environment could thus have
two distinct functions: either to remove limitations in the availability of essential metals or to raise the supply to toxic levels.

Pollution by domestic and industrial sewage contributes four
main classes of substances to the metal balance and its ecological
expression: (1) the metals, (2) nutrients for phototrophic and
(3) heterotrophic organisms, and (4) organic substances capable of
interacting with metals and thus able to regulate their bioavailability.

In experimental organisms studied under defined laboratory
conditions the transition from deficiency to toxicity for a specific
metal occurs within a relatively narrow range of about 2 orders of
magnitude. Many external and internal factors modify the reactions
of organisms toward metals; consequently a concentration which in
one situation is deficient may become toxic as the conditions change.
Therefore the recorded ranges of deficiency and toxicity may easily
overlap. The aim of this contribution is to discuss mechanisms
which influence the response of aquatic communities enriched with
essential metals.

2. ESSENTIAL METALS

2.1. Requirements for Essential Metals

The demonstration of a physiological requirement for an element in
any animal or plant depends on the ability of that species to grow
and reproduce while receiving inorganic nutrients in dissolved or
solid form of a clearly defined composition. Our knowledge of plant
and animal nutrition is derived predominantly from the study of agriculture and of some laboratory studies, and of course studies on
humans. Agriculture has sought improved productivities at high
nutrient levels, and organisms from nutrient-deficient environments
have received little attention in the study of nutrient requirements.

Essential metals are micronutrients required in small amounts.
For example, a low requirement for cobalt is suggested by the fact
that only three molecules of vitamin B_{12} per cubic micrometer of
cell is considered sufficient for the growth of *Euglena* [1]. Assuming a cobalt requirement of 1 atom per cubic micrometer of cell,
this represents a tissue concentration of 10^{15} atoms/dm^3, or 10^{-8} M.
If organisms concentrate cobalt 1000-fold above the external level,
concentrations as low as 10^{-11} M (around 0.5 ng/liter) would ensure
an adequate supply. Therefore the demonstration of metal requirement
calls for highly sensitive analytical techniques. Studies on trace
nutrient require the use of ultrapure chemical and the total exclusion of contamination from water, glassware, and dust; but even so
experimental organisms often contain sufficient reserves of trace
nutrients to support significant growth.

Many groups of ecologically important organisms like fish,
amphibia, reptiles, insects, and annelids have been virtually
ignored with respect to their essential metal nutrient requirements.
For these groups requirements are primarily inferred from the presence of enzymes with particular metal dependency.

The present list of mineral nutrients essential for plants and
animals comprises 23 elements (Table 1). As yet there has been no
indication of fundamental differences in metal requirements on the
basis of either plant- or animal-specific biochemical mechanisms,

TABLE 1

Essential Mineral Nutrients in Plants and Animals

Element[a]	Symbol	Atomic number	Required in plants[b]	Required in animals[b]
Nitrogen	N	7	++	++
Sulfur	S	16	++	++
Selenium (3)	Se	34	-	+
Phosphorus	P	15	++	++
Arsenic	As	33	-	+
Silicium	Si	14	(++), ±	±
Fluorine (2)	F	9	-	+
Chlorine	Cl	17	-	+
Iodine (3)	I	53	-	+
Sodium	Na	11	(+)	++
Magnesium	Mg	12	++	++
Calcium	Ca	20	++	++
Boron	B	5	+	±
Vanadium	V	23	±	±
Chromium (3)	Cr	24	-	±
Manganese (2)	Mn	25	+	+
Iron (1)	Fe	26	+/++	+/++
Cobalt	Co	27	±	±
Nickel	Ni	28	(±)	±
Copper (2)	Cu	29	+	+
Zinc (1)	Zn	30	+	+
Molybdenum (3)	Mo	42	±	±
Tin	Sn	50	-	±

[a]Requirement in humans [4]: (1) 10-20 mg/day, (2) 1-5 mg/day, (3) below 0.2 mg/day.
[b]Requirement: ++, macronutrient; +, micronutrient; ±, required in very small quantities; ( ), required only by some specialists.
Source: Compiled from Refs. 2-5.

with differences only restricted to the unique requirement of specialist species, for example, organisms able to fix atmospheric nitrogen. More elements are considered essential for animals than for plants, although this may be primarily the result of the more advanced methods used in animal studies in ultraclean laboratory environments [6].

Although considerable amounts of metals are taken up in ionic form, they are transformed into complex species within organisms. For example, 4-5 g of iron is normally present in a 70-kg man; 70.5% is found in hemoglobin, 3.2% in myoglobin, 26% in storage compounds (ferritin and hemosiderin), 0.1% in transferrin, 0.1% in cytochromes, and 0.1% in catalase [3].

Metal ions are needed to ensure the function of a number of biomacromolecules [7]. The biochemical functions of the essential metals have been extensively reviewed [2,5,8] and some of the key functions of essential metals are summarized in Table 2.

Requirement should be seen as an interdependence between nutrients. There is no evidence supporting the view of a unique optimum nutrient composition for any single species of organism [9]. The interaction of copper, zinc, and silicic acid on the growth of diatoms (discussed in Sec. 4) is a good illustration of this interdependence.

Some decades ago the requirements for trace elements in algae were found to be less than 10^{-7} M (Table 3). Elements such as iron and manganese are needed in greater amounts for phototrophic metabolism than during heterotrophic metabolism by phototrophs. According to recent observations, physiological limitations may occur at concentrations lower than previously expected. The very great sensitivity to some essential elements (e.g., 10^{-10} M copper seriously inhibits photosynthesis in algae [16]) suggests that deficiencies in aquatic plants will only become evident at extremely low concentrations. In land plants deficiency also only becomes apparent at an estimated 10^{-9} M for wheat and 10^{-13}-10^{-14} M for maize [17].

TABLE 2

Some Functions of the Essential Heavy Metals[a]

Atomic number	Element	Functions
23	Vanadium	Nitrogen fixation, redox catalyst in lipid metabolism, iron metabolism
24	Chromium	In animals cofactor for insulin (glucose tolerance factor)
25	Manganese	Redox reactions, photosystem 2 in photosynthesis, lipid metabolism in diatoms, mucopolysaccharide and cartilage synthesis
26	Iron	Reversible Fe(II)/Fe(III) reactions fundamental for many biological processes, oxygen metabolism; in terminal oxidases, oxidases, peroxidases; required for porphyrin synthesis, in hemoglobin, myoglobin
27	Cobalt	Component of vitamin B_{12}, required for methylation, nitrogen fixation in blue green algae
28	Nickel	In urease; stabilizes RNA and DNA and structure of ribosomes
29	Copper	Chloroplast redox system (plastocyanin); ascorbic acid and polyphenol oxidase, involved in metabolism of phenolic compounds; oxygen carriers in snails, involved in crosslinking reactions in collagens and in pigment formation
30	Zinc	Total of 70 zinc-containing enzymes known, including carbonic acid anhydrase, dehydrogenases, alkaline phosphatase, involved in silicic acid uptake and in nucleic acid metabolism and cell division
42	Molybdenum	In nitrate reductase, aldehyde oxidase, antagonist to copper
50	Tin	Functions not known

[a]For plants see Refs. 5 and 8; for animals see Ref. 2.

TABLE 3

Requirements for Essential Metals in Unicellular Algae

Element	Species	Activity	Concentration of total metal (M)	Reference
Iron	*Chlorella pyrenoidosa*	Critical concentration		
		Heterotrophic growth	$1 \cdot 10^{-9}$	10
		Phototrophic growth	$1.8 \cdot 10^{-5}$	10
		No growth	$< 1.8 \cdot 10^{-7}$	10
	Chlorella sp.	Adequate growth	$1.8 \cdot 10^{-7} - 2.6 \cdot 10^{-8}$	12
Manganese	*Chlorella pyrenoidosa*	Good growth		
		Phototrophic	10^{-7}	11
		Deficiency heterotrophic	10^{-10}	11
Copper	*Chlorella* sp.	Deficiency	$< 10^{-7}$	12
	Scenedesmus acutus	Optimal growth	$10^{-7} - 10^{-6}$	13
Zinc	*Chlorella pyrenoidosa*	Deficiency symptoms	$< 10^{-7}$	12
	Thalassiosira weisflogii	Limitation	10^{-11a}	14
	Thalassiosira pseudonana	Full cell growth	$10^{-12.1a}$	14
Molybdenum	*Chlorella* sp.	Growth with nitrate as N source; deficiency	10^{-9}	15

[a]The concentration refers to free metal in ionic form.

2.2. Deficiency in Natural Environment

The concentrations of essential metals in natural waters are on the
order of a few micrograms per liter (10^{-8}-10^{-7} M; see Table 4).
However, rivers subjected to mining pollution typically have metal
concentrations some 2 orders of magnitude greater [22]. Factors
and processes that regulate metal concentrations in aquatic environments have been discussed recently [23]. For example, in Swiss
lakes, about one-half of the input of zinc and copper and one-tenth
of iron results from atmospheric deposits [24].

The effect of metal deficiency on the productivity in aquatic
environments has been studied predominantly on oligotrophic lakes
[25], and of the 28 oligotrophic lakes investigated in different
parts of the world a significant deficiency in trace metals was
found in 82%.

In Blelham Tarn and Esthwaite Water (Great Britain) a requirement for iron was noticed, but after the addition of either Fe(III)-
EDTA or EDTA alone, phosphorus limitation occurred [26]. The effects
of metal additions are influenced by multiple interactions between
metals, organic substances and sorptive surfaces and are difficult
to interpret (see Sec. 6.3). Results on iron limitation for phytoplankton are inconclusive, but the possibility remains that growth
rates are dependent upon specific reactions involving chelates and
the cells [27].

The role of essential metals in limiting productivity in lakes
has been intensively discussed [25], and since phosphorus was recognized as the main limitation of productivity of lakes in temperate
regions metal deficiency has received less attention [28].

The difficulties in providing a nutrient supply of known composition have largely precluded deficiency studies in aquatic animals other than in fish. In fingerling rainbow trout (*Salmo gairdneri*) an albumin-based diet containing zinc at 4 μg/g resulted in
poor growth and high mortality, whereas addition of zinc at 90 μg/g
reduced mortalities and improved growth. The trout showed no sign

TABLE 4

Concentrations of Some Essential Heavy Metals in Swiss Waters (in μg/liter)

	V (at. no. 23)	Mn (at. no. 25)	Fe (at. no. 26)	Co (at. no. 27)	Cu (at. no. 29)	Zn (at. no. 30)	Mo (at. no. 42)
Groundwater							
Zürich/Tüffenwies[a]	—		13.7	2	2.8	5.8	1
Zürich/Tüffenwies[b]	<3	<1	<3	0.03	1.5	5.7	1.6
Rivers							
Glatt/Fällanden[a]			50			10	
Glatt/Glattfelden[c]			100			30	
Aare[d]			100		2	20	
Reuss[d]			130		2	20	
Rhein[d]			110		2	10	
Rhone[d]			330		1	10	
Lakes							
Lake Lucerne[e]			25		8	10	
Greifensee[e]			45		5	18	

[a] Data from the EAWAG; monthly average for 1980-1981 (filtered by a 0.45-μm filter).
[b] Data from Ref. 18.
[c] Data from Ref. 19.
[d] Data from Ref. 20.
[d] Data from Ref. 21.

of Zn toxicity during feeding with levels of dietary Zn as high as
1700 µg/g [29].

Terrestrial organisms depend on the local geological strata
for their nutrients. However, aquatic habitats receive their mineral supply from the entire upstream area, increasing the likelihood
of a balanced chemical composition. Consequently, direct metal
limitations resulting from a deficient supply may occur less frequently in aquatic environments.

3. UPTAKE AND INCORPORATION OF METALS

3.1. Uptake of Metals

Most aquatic organisms resorb metals from solutions over their
entire surface; animals incorporate them additionally from particulate food. In submerged plants roots are of minor importance for
nutrition, as minerals enter through leaves [30].

Uptake of metals depends on maintenance of concentration
gradients across cell membranes. These are established by the
accumulation of ligands binding the metals on the cell surface,
transfer through cell membranes by carrier ligands, and removal of
metal ions inside the cell by special biomolecules. Metabolic
activity is responsible for the synthesis of different ligands [31].

The rate of uptake in animals and plants for copper and zinc
and the toxicity of these metals to animals and plants is normally
proportional to the free ion concentration and not the total metal
concentration [32]. Complexed forms are in general not taken up;
however, there is still some disagreement about the fate of bound
metals at cell surfaces and the degradation of organic chelators.

The mechanism of uptake is fast [33] and efficient. In general, the ratio of the concentration of elements inside an organism
to that outside, the so-called *concentration factor* (CF) is on the
order of 10^3-10^4. The actual value depends on the metal and its

speciation, the organism, and environmental conditions and varies within considerable limits (about 2-3 orders of magnitude).

The special conditions for the uptake of substances in surface films at the air-water interface has been summarized recently [34].

Chelation and the formation of protein compounds transform the generally toxic ionic metals to nontoxic bound forms suitable for uptake, transport, and storage [35]. Compounds characteristic of the different functions include

Siderochromes for the uptake of metals by sequestration

Intracellular transport proteins

Storage proteins like metallothioneins

The half-life of these proteins is on the order of a few days, so the maintenance of metal homeostasis is a dynamic process. Simpler compounds, especially organic acids, are involved in the metal transport in plants.

3.2. Relative Importance of Food and Ambient Water for Metal Uptake

For aquatic animals food is in general a more important source of metals than water [36]. In a simple marine food chain with the brown alga *Fucus serratus* the herbivore snail *Littorina obtusa* assimilated 1.1 µg/g Zn per day from food compared to 0.38 from water [37]. In a model food chain utilizing labeled ^{65}Zn, the alga *Clamydomonas* sp. were fed to *Artemia* sp., which in turn was offered to two species of fish, *Gambusia affinis* and *Leiostomus xanthurus*, in a heavy metal chelate buffer system maintained at a stable Zn ion activity ($10^{-8.5}$ M) [38]. Food provided about 80% of the total accumulation of ^{65}Zn by the fish.

Metals in the food may be in an unavailable form. Zinc tends to be nonexchangeable when accumulated. In juvenile mosquito fish (*G. affinis*) three zinc pools were detected, the largest with a half-life greater than 230 days [39].

3.3. Role of Chelators Secreted by Aquatic Organisms

Natural chelators in lake water are of great importance as regulators of the concentration of free metals [40]. For example,
polyphenol containing exudates of brown algae (*Fucus* and *Ascophyllum*)
protect the heavy-metal-sensitive *Sceletonema costatum* from Zn^{2+}
(0.5-2 mg/liter) [41]. Similarly, the crustacean *Daphnia magna*
releases agents complexing copper at an equivalent rate of $40 \cdot 10^{-12}$
M per hour and animal, while with a density of 1 animal/ml, 10^{-6} M
Cu could be complexed per day. This ability may affect the result
of toxicity tests [42].

     Frequently metals outside an organism are not in a readily
available form. Several mechanisms exist that mobilize bound metals.
Blue-green algae can excrete iron-selective chelators, giving them
a competitive advantage over other algae [43]. *Anabaena cylindrica*
can produce about twice as much complexing ligand than *Navicula
pelliculosa* and 10 times as much as *Scenedesmus quadricauda* [44].
As a result of the excretion of chelators, it is likely that some
fraction of particulate metals, which are normally excluded from
analysis by filtration through 0.45-μm pore filters, become bioavailable.

4. TOXICITY AS A RESULT OF OVERSUPPLY
OF ESSENTIAL METALS

4.1. Mechanisms of Metal Toxicity

"Toxicity" in an ecological context refers to chemical effects by
substances which reduce the fitness of single populations and alter
interactions between populations. Initial interest in studying
toxicity centered on lethal effects. However, to understand the
long-term consequences of pollution in ecosystems it is essential
to recognize sublethal effects. These can be classed into (1)
morphological changes, (2) effects on growth rates, sexual development, and reproduction rates, (3) behavioral changes, which result

in lower ability to escape predators or compete effectively, and
(4) genetic modifications. The vast amount of information on the
toxicity of essential metals has been reviewed in several literature surveys [45,46].

The cause for toxicity rests in a general way on the interaction between metals and bioactive proteins and involves mechanisms
similar to those responsible for the essentiality of metals. Deleterious effects caused by metal ions in biological macromolecules
comprise the following:

Displacement of an essential metal from an active site by
        a "toxic" metal
Binding to undesired parts of a macromolecule
Crosslinking, which can produce undesired aggregates
Depolymerization of biological molecules
Misrepair of nucleotide bases and induction of errors in
        protein synthesis [7]

At present no possibility exists to assess the importance of
any of these potential mechanisms for in vivo toxicity [7]. Although
a general theory presenting metal toxicity under unifying principles
is yet to come, some theoretical models have been proposed in attempts
to explain toxic effects in fish [47].

4.2. Range of Susceptibility to Essential Metals

One of the most striking aspects about compilations of toxicity data
is the wide range of the values reported (Table 5). Factors that
regulate availability, uptake, and effects are of varied nature and
can be attributed to (1) chemical factors (speciation, accompanying
substance, redox potential, and rhythm of exposure), (2) physical
factors (light, temperature, and turbulence), and (3) biological
factors (size of organisms, developmental stage, nutritional status
and general health conditions, acclimation, and feeding strategies).

In animal orders studied more intensively the sensitivity
toward metals expressed in LC_{50} (median lethal concentration,

TABLE 5

Range of Acute Sensitivity of Major Groups of Fresh-Water
Organisms to Toxic Concentrations (in mg/liter) of
of Copper and Zinc[a]

	Copper	Zinc
All test organisms	0.005->100	0.04-200
	>20,000	5000
	n = 209	n = 94
Crustaceans	0.005-9.6	0.04-19
	1900	480
	n = 31	n = 8
Insects	0.031->100	1.6-32
	>3200	20
	n = 12	n = 5
Fish	0.015->100	0.093-200
	>6700	2100
	n = 126	n = 71
Fish (chronic test)	0.05-0.12	0.51-1.4
	24	27
	n = 20	n = 27
Other organisms	0.006->100	0.43-16
	>16,000	37
	n = 40	n = 10

[a]Concentrations are the LC_{50} values for 1 day or longer. The chronic
fish test gives the lowest concentration of metal having a harmful
effect in chronic tests. Each range is followed by the ratio of the
highest concentration to the lowest, n being the number of tests.
Source: Data extracted from Ref. 48, Table 1.

killing 50% of a population in a given time) varies over 2-4 orders
of magnitude. The most sensitive animals are 1000-10,000 times more
susceptible than the most resistant ones. As more data accumulate,
it becomes evident that the variation in susceptibility within
species cannot be attributed to test conditions alone, but is inherent to organisms [48]. The contribution of genetic variation has
received considerable attention in plants [49].

When the range of deficiency and toxicity are compared in a
particular algal strain, they are separated by about 1.5-2 orders of
magnitude. In the dinoflagellate Gonyaulax tamarensis copper limita-

tion occurs at $5 \cdot 10^{-13}$ M Cu activity, with an optimum around 10^{-12} M and an abrupt drop to no growth at 10^{-11} M [50]. If an organism is studied at different times, the level for cardinal values such as the LC_{50} varies within considerable limits. Sensitivity to copper (as LC_{50}) in natural populations of the copepod *Acartia tonsa* were found to vary over a period of 6 months between 100 and 300 ppb copper [51].

An extensive review of the experiments performed in different laboratories over the world under comparable conditions (flow through system with a temperature range of the water of 12-15°C) shows that rainbow trout (*Salmo gairdneri*) are the most sensitive as well as the most resistant fish [48]. In hard water the range of the LC_{50} toward copper is ninefold (94-870 μg/liter Cu) and in soft water the range is fourfold (15-54 μg/liter). This sensitivity range found in rainbow trout is attributed to biological variations in the fish used for experiments and does not differ significantly from the range observed in all the other fish species taken together. Considerable ranges in sensitivity have been observed under carefully controlled conditions in the same laboratory, with a fivefold variability over a period of several months [52].

On the whole there is little agreement on the relative sensitivity of major taxonomic groups of fresh-water animals. Salmonids have been claimed to be a relatively sensitive group of fish. Cladocerans and amphipods and mayfly nymphs appear to be relatively sensitive among crustaceans and insects; but good comparative data linking these groups are lacking. Indication for relative sensitivity comes from experiments with mixed communities (see Sec. 6); however, shifts in the composition of communities may result from indirect causes and can therefore not be attributed to direct effects of the toxicant without appropriate investigations. For reliable generalizations a wide variety of animals have to be tested under stable and similar experimental conditions [48]. In comparing the published data on copper toxicity in a comprehensive search Skidmore and Firth conclude the following [48]:

The ranges of sensibility of major taxonomic groups of freshwater organisms to heavy metal toxicity show almost
total overlap.

The available data indicate no general order of sensitivity
at either the species level or higher taxonomic levels.

4.3. Functional Expressions of Toxicity

Compared with metals in general, the toxicity of essential metals
ranks high (Table 6). In aquatic organisms only mercury is consistently more toxic than copper. Copper has received more attention
than any other metal in aquatic environments; its high toxicity and
comparatively high concentration in water make it a critical element.
In order to illustrate the multiplicity of deleterious effects by
metals some of the effects of copper on specific biological functions
will be presented.

Nitrogen fixation: Toxic effects of copper have been reported
in blue-green algae at concentrations of about 10^{-7} M
[57]. Inhibition of N_2 fixation was identified as a
cause of this sensitivity.

Photosynthesis· In algae copper inhibits photosynthesis at
very low concentrations (10^{-10} M ionic Cu) [16].

Silicic acid metabolism: In short-term exposures of marine
phytoplankton to 2.5, 5, 10, and 25 µg/liter Cu the
silicic acid uptake was found to be inhibited by copper
[58]. Copper toxicity is a function of silicic acid
concentration in the range of 10^{-5}-10^{-4} M Si(OH)$_4$ [59].
It has been shown that zinc limits silicic acid uptake,
copper competing with zinc for the active sites. Transport of copper into the cells is a function of the ratio
of Zn to Cu activities [60].

Growth: The copper-to-zinc ratio determines growth response
in *Euglena* [61]. A 50% growth depression in the marine
diatom *Nitzschia closterium* was observed at a concentra-

TABLE 6

Toxicity Sequence of Metals

Organism	Sequence[a]	Reference
Animals (LC_{50})	Hg > Ag > Cu > Zn > Ni > Pb > Cd > As > Cr > Sn > Fe > Mn > Al > Be > Li	53
Insects	Cd > Hg > Cu >> Zn > $\underline{Cr}$ > Pb > $\underline{Ni}$	48
Crustacea	Hg > Cd > Cu > Zn > Pb > Co > Ni	48
Fish	Cd > Hg > Cu > Zn > Pb > $\underline{Ni}$ > $\underline{Cr}$	48
Phytoplankton, meso- and eutrophic Swiss lakes	Hg > Cu > Cd > Zn > Pb	54
Heterotrophic plankton glucose uptake	Hg > Cu > Cd > (Zn, Pb)	55
Diatoms (*Nitzschia closterium*, EC_{50} growth)	Cu > Zn > Co > $\underline{Mn}$	56

[a]Elements are underlined when the toxicity is in the range of milligrams per liter (for Refs. 48 and 56).

tion of 33 µg/liter copper, 270 µg/liter zinc, 12,000
µg/liter cobalt, and 25,700 µg/liter manganese [56].
Growth has been observed to be more susceptible to
copper than photosynthesis in *N. closterium* [62].

Feeding activity: Sublethal effects on feeding activity and
egg production in the marine copepod *A. tonsa* were
noticeable during a 4-day exposure at 10^{-7} M Cu. This
is about 1-1.5 orders of magnitude lower than the LC_{50}
for 24 hr. Feeding intensity in several marine macro-
zooplankton was reduced at 5 and 10 µg/liter Cu [63].

Reproduction: The sensitivity of reproduction toward copper
is similar to feeding activity and is affected by 10^{-7} M
copper [63]. Egg development in laboratory-tested *D.*
longispina and *Eudiaptomus gracilis* grown in limnocorals
with a metal mixture was reduced [64] (see Sec. 6).

Behavior: In hard water 125 µg/liter Cu affects the capacity
of larvae of *Hydropsyche angustipennis* to spin nets.
The primary site of action seems to be in the silk gland
[65]. Copper at 8 µg/liter depresses the olfactory
response of fish toward the standard stimulant L-serine
[66].

4.4. Environmental Factors Affecting Toxicity

The main physicochemical factors affecting the response of an organism to toxicants are temperature, dissolved oxygen, pH, hardness,
alkalinity, and the presence of chelating agents and other pollutants,
and these have been extensively reviewed [67]. In general, sensitivity to metals is reduced with increasing pH and hardness of the
water and decreased when oxygen partial pressure falls. Age, size,
nutritional status, and acclimation are also important variables
determining the sensitivity of animals and plants. In general,
larger animals are more resistant than smaller ones. For chaeto-
gnaths (*Sagitta hispida*), with biomasses ranging over 3 orders of

magnitude, the logarithm of LC_{50} toward copper concentrations lies between 40 and 400 µg/liter and is proportional to the logarithm of the biomass. A similar relationship was found between 13 marine species which could be grouped into high-, average-, and low-sensitivity categories when compared at the same body weight [63].

Nutritional status is important. A diet rich in vitamin C or with casein and gelatin as a protein source instead of fish and soybean meal conferred in rainbow trout a higher tolerance to chlordane [68]. In the calanoid copepod *Acartia tonsa* the LC_{50} for 72 hr toward copper ranges between 9 and 78 µg/liter and depends on the population density and decreases with increasing food availability [69].

In a comparative study on the effect of temperature on the action of several toxicants on plankton algae, zooplankton, microbenthos, macrobenthos, and fish, the response was nonuniform [70]. For example, in fish a "crossover" was observed, that is, with copper below 4 mg/liter bluegill sunfish survived longer at 30°C, whereas with copper above 4 mg/liter they survived longer at 15°C.

Synergism and antagonism between heavy metals have been widely studied [54]. Iron has a marked effect on the toxicity of copper for *Chlorella* [33]. The effect of iron is suspected to result from the scavenging of other metals in the water and will be discussed in Sec. 6.

Young steelhead trout (*Salmo gairdneri*) exposed to copper over 78 days accumulated less copper and survived better than animals intermittently subjected to the same daily average concentration. This suggests that data derived from continuous laboratory exposure should not be directly converted into water quality criteria, where toxicant concentrations in natural water are inevitably variable [71].

5. TOLERANCE TO METALS

5.1. Tolerance in the Field

Algae, predominantly chlorophytes (*Hormidium, Microspora, Stigeoclonium,* and *Ulothrix* species) [22], tolerant to high metal concen-

trations in the range of milligrams per liter are found in rivers
running through mining districts. In enclosures at much lower metal
concentrations than those in mining-polluted rivers (about 2-3 orders
of magnitude) planktonic communities react to metal stress by a change
in species composition toward less susceptible species [72,73].

5.2. Mechanisms of Tolerance

Tolerance toward metals can be achieved by (1) reduction of uptake,
(2) increased excretion, or (3) transformation into an inactive form
by either sequestration or precipitation [35].

Slime may serve a protective function against metals by reducing diffusion rates to the cell surfaces. The alga *Mesothaenium*
(Zygnemales) produces a water-soluble slime consisting of 30-40%
uronic acid residues capable of binding copper. Algae producing
more slime were more resistant to Cu, their slime containing more
Cu and the cells showing less Cu accumulation [74].

Reduced metal uptake rates by some unknown mechanisms seem to
be responsible for the metal tolerance of phytoplankton in enclosures
[72]. Immobilization of metals can result from calcium deposits,
both intracellular and extracellular, that sequester divalent cations
in a number of marine invertebrates. Lysosomes, membrane-bound cytoplasmic bodies of 0.25-0.5 μm diameter with hydrolytic enzymes, compartmentalize cations. These protective techniques allow high metal
concentrations which may be transferred to the next trophic level
[35].

The synthesis of metallothioneins is induced by several metals,
including Zn, Cu, Cd, Ag, and Hg. Induction by one metal may confer
resistance to another because of the nonspecific nature of ligands
[75].

As a generalization it appears that the plants and the bivalve
molluscs are poor regulators of heavy metals; the essential metals
Zn and Cu, on the other hand, are regulated in decapod crustaceans
and fish [36].

6. EFFECT OF ESSENTIAL METALS ON AQUATIC COMMUNITIES

6.1. Introduction

Domestic and industrial sewage contributes metals to aquatic ecosystems in association with other sewage compounds. As a result,
changes in the aquatic environment are difficult to interpret with
respect to the contribution of metals. Therefore the effects of
metals on the composition of aquatic communities have been studied
in special environments on several levels:

> In laboratory systems with artificial assemblages of a few
>   selected species, mostly representing simple food chains
> In large-scale experimental systems, in enclosures for the
>   study of limnic and marine planktonic communities, and
>   in artificial rivers for the study of benthic commu-
>   nities in lotic environments
> In natural waters, especially rivers, subject to higher loads
>   of metals, either naturally or by mining activities

In the field loading with metals occurs in rivers flowing
through metal-rich geological strata or in mining areas. The long-
term exposures lead to a selection of metal-resistant ecotypes. In
general, the pollution in mining areas is composite; it includes
several metals simultaneously at high concentrations (mg/liter) and
low pH values as a result of the oxidation of sulfides. Floods
cause intensive abrasion due to the action of suspended mining grit.
For this reason mining causes extreme forms of pollution and will
not be discussed here. A review on lotic rivers in mining areas is
given in Ref. 22.

6.2. Laboratory Studies in Simple Food Chains

The transfer of specific substances along food chains has been
reviewed for lipophilic substances [76], ionic and polar compounds,

including metals [77], and metals in invertebrates [78]. One conclusion is that metals do not accumulate in the edible portions of
fish, except mercury and arsenic, which are found in muscle, and
cadmium, arsenic, mercury, and silver, which have been found to
accumulate in shellfish. This implies that increases in metal concentration along the food chain, that is, bioaccumulation or biomagnification, does not generally occur.

When one organism eats another, the amount of metal taken up
depends on (1) the content of the parts consumed, (2) the efficiency
of the incorporation, and (3) the intensity of excretory processes.

The grazing relation between a flatfish (*Pleuronectes platessa,*
plaice) and a bivalve (*Tellina tenuis*) is a good illustration of
metal storage in body parts which are not eaten. In natural environments the flatfish graze on the siphon of the bivalve; the siphons
are subsequently regenerated and play a significant role in the food
chain. In a laboratory system composed of young flatfish feeding on
the bivalve mollusc which lives on phytoplankton [79] copper added
at 10 μg/liter had an adverse effect on the food chain, reducing
photosynthetic primary production and growth of both bivalve and
fish. The copper was preferentially accumulated in the hepato-
pancreas of the mussel. By eating the muscular siphon, the fish
ingests little copper from its food. The highest level of copper
was found in *Tellina.* However, no bioaccumulation was observed.

Plankton algae are taken up whole by filter-feeding bivalves.
An artificial marine food chain consisting of the planktonic micro-
alga *Diognes* and the mussel *Mytilus edulis* was treated with the
metals copper, zinc, chromium, lead, and mercury (Cu, Zn, Cr, Pb,
and Hg) [80]. The concentration factors (CF) for phytoplankton
showed the sequence Zn > Pb > Hg > Cu > Cr (CF = 100-7000), and for
the mussels relative to the ambient water Zn > Hg > Pb > Cu > Cr
(CF = 40-1200). Compared to the lower trophic level, the concentrations in mussel decreased and there was no biomagnification.

The outcome of feeding experiments depends on the food-to-
biomass ratio and the duration of the experiment. For this reason

observation in the field where steady states may be expected is informative. The river Lech in Bavaria (West Germany) [81] receives the outfall of the sewage works of the city of Augsburg and is mildly polluted by metals. The largest accumulation of metals was not immediately below the metal discharge; occasionally it occurred above the point of entry, especially in insect larvae, as a result of migration. Accumulation was clearly greater at the base of the food chain (in sewage fungus and macrophytes) than in detritivores and herbivores, *Carinogammarus*, and the predatory leech *Erpobdella* (Table 7). Of the metals added, namely, cadmium, mercury, copper, and lead, copper was most easily taken up.

In confirmation of the generally observed ability of higher trophic levels to prevent excessive accumulation of metals, no increase in metal content was found in the higher trophic levels in a limnic enclosure [82].

TABLE 7

Bioaccumulation in Selected Organisms in
the River Lech (Bavaria)[a]

	Cd	Cu	Hg	Pb
Typical in German rivers (µg/liter)	<0.1-0.5	<1-7	<0.1	<0.5-3
Lech, average	≈2-≈1	14-≈0.5	≈0.6-0.1	20-3
In organisms (ppm dry weight)				
Sphaerotilus natans	4.5-<0.5	125-21.8	8.1-<0.1	55-<19.9
Ranunculus fluitans	13.8-1.5	90.6-47.5	22-0.6	31.7-<9.0
Carinogammarus	3.3-<0.7	28.2-16.8	<0.8-<0.2	18.0-<2.7
Erpobdella	<0.7-<0.2	9.6-5.1	<0.3-<0.2	11.4-5.6

[a]Given are the values at the stations with the highest and lowest metal concentrations along the river Lech below the sewage inflow of the city of Augsburg.
Source: Data extracted from Tables 69 and 70 in Ref. 81.

6.3. Reactions in Complex Seminatural Communities

The effect of a mixture of the essential and nonessential metals
zinc, copper, cadmium, lead, and mercury at the Swiss tolerance
limits (in 10^{-8} M: Zn, 308; Cu, 16; Cd, 4.5; Pb, 24; and Hg, 0.5)
was studied in enclosures in the strongly eutrophic Swiss lake
Baldegg. The load adversely affected phytoplankton, zooplankton,
and the bottom fauna [72]. The productivity in the loaded containers was only significantly lower than in the controls the first
year. At the beginning of the second year productivity in the polluted system rose and exceeded the control several-fold as a result
of changes in phytoplankton composition to more metal-resistant
species.

The grazer population was damaged also; the density of crustacean plankton was on the average 2.5 times higher in the control
than in the loaded limnocorrals [64]. Egg development of the species
collected in the metal-loaded environment was slower than in control
animals, suggesting that reduction of populations densities was a
direct effect of metals on reproduction and could not be attributed
indirectly to changed food supply.

Fish fry kept in water piped from the enclosures seemed able
to control their copper and zinc content and to maintain it at a
constant level, regardless of environmental concentrations [82].

On a dry weight basis, metal concentrations were higher in
phytoplankton and periphyton samples, lower in zooplankton, chiron-
midae and *Sialis*, and lowest in fish, indicating that inorganic Hg,
Cu, Cd, Zn, and Pb did not accumulate through the food chain and
did not reach highest concentrations in the highest trophic level.

In the metal-loaded container a higher content of amino acids
and a shift in their composition was found. This suggests the possibility of regulation of metal uptake or metal toxicity by biological reactions in the stressed enclosures. Dissolved macromolecules and colloidal substances present in lake water were shown
to complex copper and reduce its toxicity [40].

In marine enclosures at Saanich Inlet (Vancouver Island, Canada) the effect of copper alone was investigated. Addition of 10, 20, and 50 µg/liter of copper caused initial inhibition of phytoplankton crop, photosynthesis, and growth rates, followed by a recovery [83]. At the end of 27 days productivity values were the same in treated and untreated enclosures, but the species composition had changed. In the copper-treated enclosures the initially present centric diatom *Chaetoceras* was replaced by microflagellates and some other diatoms. When in a later experiment copper-resistant species were already present, the addition of 5 and 10 µg/liter Cu seemed to increase standing crop over the control level, but not the rate of photosynthesis, however. Copper reduced diversity in the first experiment but not in the second one.

Zooplankton in the presence of 5, 10, and 50 µg/liter Cu [84] showed changes in species composition over the whole concentration range. Animals collected in enclosures to which 5 and 10 µg/liter Cu had been added all showed reduced feeding activities in the laboratory compared to controls [63].

At the level of 10 and 50 µg/liter Cu the relative number and activity of bacteria increased markedly. This was probably the result of the release of organic substrates from the copper-sensitive organisms in the original planktonic community [85].

In running water studies the effects of the three essential elements Co, Cu, and Zn were investigated in artificial outdoor rivers at the Swiss tolerance limit (Co, 50; Cu, 10; Zn, 200 µg/liter, respectively, or $85 \cdot 10^{-6}$, $16 \cdot 10^{-6}$, $306 \cdot 10^{-6}$M). In established spring and early summer communities the dominant alga, *Diatoma hiemale*, was destroyed by the full-strength mixture within 3 days. At only half the metal concentration colonization of the channels by *D. hiemale* was prevented. In the second month of metal stress a new community of filamentous algae dominated by *Hormidium* sp. emerged and attained production rates similar to the control (about $2g$ carbon/m^2 day).

When dosing of metals was discontinued, *D. hiemale* returned and formed mixed communities with the green filamentous algae in

contrast to the control, which was dominated only by D. *hiemale*.
The imprint of previous metal stress was apparent for many months
[86].

The reaction to metals in toxic concentrations is very similar
in all three types of communities, marine, limnic, and lotic. When
stressed with metals in the lower toxicity range, benthic running
water communities also display the basic pattern observed in planktonic communities [87].

The reaction to metal stress involves three steps:

1. Metal-sensitive forms on all trophic levels are eliminated.
2. Species adapted to the metal stress take over the newly
   available niches.
3. Reconstruction of the community takes place as a result of
   the interaction between the trophic levels under the new
   conditions.

The response of the system will depend on the size and diversity
of the pool of potential colonizers. In natural systems there will be
a great diversity of starting conditions on the biological level, and
as a consequence system responses may be variable in successive experiments [83]. The productivity level of the new community depends on
whether some metal-resistant organisms are able to exploit resources
more efficiently or survive additional stress more successfully.

In all the three types of communities studied, marine, limnic,
and lotic, the original productivity was reestablished after a tran-
sitory depression, so that after some time the toxic effect could
only be recognized in the changed composition of the communities.
If the history of the habitat were not known, the toxic effect may
not even have been detected. Sensible definitions of what will have
to be considered toxic for communities must distinguish between these
two aspects of toxicity, the effects on productivity and the effects
on sociology. As actual communities contain different potential
reactions to environmental stress, the knowledge of the time course
of the reactions and the resilience to stress will become an essential part of an ecological description of toxic effects.

6.4. Interaction of Essential Metals with Iron

In a study on the effect of nitrilotriacetic acid (NTA) on metal
availability to phytoplankton, limnic enclosures were enriched with
NTA (50-9000 µg/liter) and/or iron (30-90 µg/liter) [88]. In the
oligotrophic lake of Lucerne added Fe (or NTA) caused an increase
in photosynthetic rates in the first 4 days; on the other hand, in
the eutrophic lake the same algal species were inhibited. Although
in the eutrophic Greifensee there is only one-tenth as much iron
available per unit biomass as in lake Lucerne, addition of iron
reduced productivity. Iron was therefore not seen as growth
limiting.

The iron was considered to act by scavenging the available
free metal ions. In the oligotrophic lake the amount of organic
substances able to complex trace metals is too low to reduce free
metal ion concentration to a nontoxic level. The addition of iron
therefore acts as a detoxifier by adsorbing the metals. In the
eutrophic lake, on the other hand, 6-10 times more organic compounds
are present for the concentration of copper and zinc. Organic substances compete with algae for essential metals. Addition of Fe and
NTA further enhances this competition to a point where metals become
limiting to algal growth.

In artificial running water communities the toxic action of
the essential elements Co, Cu, and Zn mentioned previously is neu-
tralized by iron as Fe-EDTA (300 µg/liter or $5.38 \cdot 10^{-6}$ M/liter Fe)
but not by EDTA alone [89]; however, the shift in the composition of
the community by the toxic metals is not reversed (Table 8). The
action of iron may be similar to that observed in the oligotrophic
lake [88] by scavenging the metals Co, Cu, and Zn to nontoxic concentrations.

The observations on artificial lotic communities suggest that
the occurrence of green algal communities in central European rivers
otherwise suitable for the growth of streamer-forming diatoms may
result from a mild but continuous metal stress.

TABLE 8

Effect of Iron on the Development of Microphytic Benthos in Channels
Loaded with the Essential Metals Co, Cu, and Zn [89]

| | Control | | Co, Cu, Zn | Co, Cu, Zn + EDTA | Co, Cu, Zn + Fe-EDTA |
	Channel 1	Channel 2	Channel 3	Channel 4	Channel 5
Metal (μg/liter)					
Fe	7		8	7	307
Co	<1		55	60	60
Cu	1		10	10	10
Zn	13		189	185	183
Productivity (g dry weight/m^2 day)					
3rd week	13.1	14.2	2.7	2.7	14.3
4th week	15.4	17.0	2.6	6.0	13.2
In percent of control	100		18	29	92
Dominance of diatoms[a]					
After 30 days	+++	+++	+	++	++
After 80 days	+++	+++	-	-	-

[a]Absolutely dominant, over 90% of the biomass, +++; dominant, over 50%, ++; present, less than 50%, +; absent, -. Concentrations (in 10^{-6} M) are Fe, 5.38; Co, 0.85; Cu, 0.16; Zn, 3.06; EDTA, 5.38.

Higher natural iron concentrations are occasionally found in small lowland rivers. In a small oligotrophic stream in Mid-Jutland (Denmark) a moderate ferric hydroxide load from drainage water (about 1 mg total Fe per liter, 6 ppm Fe in sediment) caused a reduction in species number, and the estimated gross annual production of periphyton attained 71 g O_2/m^2 in the unloaded section and 33 g O_2/m^2 in the loaded section [90].

7. CONCLUSIONS AND OUTLOOK

In aquatic ecosystems some of the essential metals, as, for example copper, are found in amounts near the transition from essential to concentrations, causing sublethal effects in seminatural communities (in the range of 1-5 µg/liter Cu, or $2 \cdot 10^{-8}$-$10 \cdot 10^{-8}$ M). In this labile situation interference with the availability of such a metal will have physiological and possibly ecological consequences.

The past years have brought an increased understanding of the role that chelation and scavenging exercise on the availability of metals. In their natural environment, especially when attached to a substrate or forming otherwise dense populations, organisms are surrounded by clouds of dissolved organic matter ready to react with metal ions and metal compounds. This will result in a highly diverse pattern of metal availability near organisms that is different from the medium from which chemical values are derived. An essential basis for understanding community reactions is the development of a "microtopic" chemistry, a chemistry describing the true environment of aquatic life. The ecologist will want to learn more about the physiological and biological reactions of selected species in different microenvironments in order to gain a better knowledge of the observed variability of responses to seemingly defined and uniform chemical conditions.

Although aquatic organisms possess at their surface no mechanism enabling them to exclude metals, biologically produced chelators

are essential for the ambient metal status. It is not yet known if
this production is adaptive or just a side effect of cell permea-
bility and lysis. In animals and plants incorporated metals up to
a certain level are rendered inoffensive by chemical immobilization;
the low resorbability of some of these compounds may be responsible
for a loss of metals on transfer along trophic chains, resulting in
a sort of "biominimation" of metals.

ACKNOWLEDGMENTS

The author wishes to thank Professor G. Hamer, head of EAWAG's
Department of Technical Biology, for his support and assistance in
preparing the manuscript; Dr. R. Gächter for his constructive comments; and T. Mason for his cheerful assistance in rewording many
of the passages.

REFERENCES

1. M. R. Droop, *Nature, 180,* 1041-1042 (1957).

2. E. Beerstecher, in *Nutrient Elements and Toxicants* (M. Rechcigl,
    ed.), S. Karger, Basel, 1977, p. 1 ff.

3. D. P. Cuthbertson, in *Nutrient Elements and Toxicants* (M. Rech-
    cigl, ed.), S. Karger, Basel, 1977, p. 26 ff.

4. W. Mertz, *Science, 213,* 1332 (1981).

5. J. C. O'Kelley, in *Algal Physiology and Biochemistry* (W. D. P.
    Steward, ed.), Blackwell, Oxford, 1974, p. 610 ff.

6. K. Schwarz, *Fed. Proc., 33,* 1748 (1974).

7. G. L. Eichhorn, in *Ecological Toxicity Research* (A. D. McIntire
    and C. F. Mill, eds.), Plenum Press, New York, 1974, p. 123 ff.

8. H. Marschner, in *Inorganic Plant Nutrition,* Vol. 15A (A. Läuchli
    and R. L. Bieleski, eds.), Springer, Berlin, 1983, p. 5 ff.

9. W. Rohde, *Mitt. Intern. Verein. Limnol., 21,* 7 (1978).

10. C. Eyster, in *Algae and Man* (D. F. Jackson, ed.), Plenum Press,
    New York, 1964, p. 86 ff.

11. J. Myers, *Annu. Rev. Microbiol., 5,* 157 (1951).

12. J. B. Walker, *Arch. Biochem. Biophys.*, *53*, 1 (1954).

13. G. Sandmann and P. Böger, in *Inorganic Plant Nutrition*, Vol 15B (A. Läuchli and R. L. Bieleski, eds.), Springer, Berlin, 1983, p. 563 ff.

14. J. G. Rueter, Jr., and F. M. M. Morel, *Limnol. Oceanogr.*, *26*, 67 (1981).

15. P. S. Ichioka and D. I. Arnon, *Physiol. Planta.*, *8*, 552 (1955).

16. R. Gächter, K. Lum-Shue-Chan, and Y. K. Chau, *Schweiz. Z. Hydrol.*, *31*, 252 (1973).

17. J. F. Loneragan, *Fortschr. Bot.*, *44*, 92 (1982).

18. A. Wyttenbach, S. Bajo, and K. Farrenkothen, *Gas, Wasser, Abwasser*, *59*, 509 (1979).

19. J. Zobrist, J. Davis, and H. R. Hegi, *Gas, Wasser, Abwasser*, *57*, 402 (1977).

20. J. Zobrist, J. Davis, and H. R. Hegi, *Gas, Wasser, Abwasser, Sonderdruck No. 784*, 1 ff. (1977).

21. H. R. Bürgi, *EAWAG News*, *2*, 1 (1973).

22. B. A. Whitton and P. J. Say, in *River Ecology* (B. A. Whitton, ed.), Blackwell, Oxford, 1975, p. 286 ff.

23. P. Baccini, in *Metal Ions in Biological Systems*, Vol. 18 (H. Sigel, ed.), Marcel Dekker, New York, 1984, p. 239 ff.

24. J. Zobrist, *Gas, Wasser, Abwasser*, *63*, 123 (1983).

25. C. R. Goldmann, in *Nutrients and Eutrophication. The Limiting-Nutrient Controversy* (G. E. Likens, ed.), American Society of Limnology and Oceanography, Spec. Symp. No. 1, Allen Press, Lawrence (Kan.), 1972, p. 21 ff.

26. J. D. Box, *Arch. Hydrobiol. Suppl.*, *67*, 81 (1983).

27. C. S. Reynolds, *The Ecology of Freshwater Phytoplankton*, Cambridge University Press, Cambridge, 1984.

28. D. W. Schindler, in *Estuaries and Nutrients* (B. J. Neilson and L. G. Cronin, eds.), Humana Press, Crescent Manor, Clifton, New Jersey, 1981, p. 71 ff.

29. J. C. Wekell, K. D. Shearer, and C. R. Houle, *Prog. Fish. Cult.*, *45*, 144 (1983).

30. Y. Waisel and M. Agami, *International Symposium on Aquatic Macrophytes*, Nijmegen, Holland, 1983, p. 287.

31. J. M. Wood, in *Metal Ions in Biological Systems*, Vol. 18 (H. Sigel, ed.), Marcel Dekker, New York, 1984, p. 223 ff.

32. U. Borgmann, in *Aquatic Toxicology* (J. O. Nriagu, ed.), John Wiley, New York, 1983, p. 73 ff.

33. E. Steeman Nielsen and L. Kamp-Nielsen, *Physiol. Planta.*, *23*, 828 (1970).

34. G. W. Fuhs, *J. Great Lakes Res., 8,* 312 (1982).

35. S. G. George, in *Physiological Mechanisms of Marine Pollutant Toxicity* (W. B. Vernberg, A. Calabrese, and P. Thurberg, eds.), Academic Press, New York, 1982, p. 3 ff.

36. G. W. Bryan, in *Effect of Pollutants on Aquatic Organisms* (A. M. P. Lockwood, ed.), Cambridge University Press, Cambridge, 1976, p. 7 ff.

37. M. L. Young, *J. Mar. Biol. Assoc. U.K., 55,* 583 (1975).

38. J. N. Willis and W. G. Sunda, *Mar. Biol. (Berlin), 80,* 273 (1984).

39. J. N. Willis and N. Y. Jones, *Health Phys., 3,* 381 (1977).

40. R. Gächter, J. S. Davis, and A. Mares, *Environ. Sci. Technol., 12,* 1416 (1978).

41. M. A. Ragan, C. M. Ragan, and A. Jensen, *J. Exp. Mar. Biol. Ecol., 44,* 261 (1980).

42. W. Fish and F. M. M. Morel, *Can. J. Fish. Aquat. Sci., 40,* 1270 (1983).

43. T. P. Murphy, D. R. S. Lean, and C. Nalewajko, *Science, 192,* 900 (1976).

44. C. M. G. Van den Berg, P. T. S. Wong, and Y. K. Chu, *J. Fish. Res. Board Can., 36,* 901 (1979).

45. A. Demayo and M. C. Taylor, *Guidelines for Surface Water Quality,* Vol. 1, Inland Waters Directorate, Water Quality Branch, Ottawa, Canada, 1981, p. 55.

46. M. C. Taylor and A. Demayo, *Guidelines for Surface Water Quality,* Vol. 1, Inland Waters Directorate, Water Quality Branch, Ottawa, Canada, 1980.

47. D. S. Szumski and D. A. Barton, in *Aquatic Toxicology and Hazard Assessment* (W. E. Bishop, R. D. Cardwell, and B. B. Heidolph, eds.), ASTM, Philadelphia, 1983, p. 42 ff.

48. J. F. Skidmore and I. C. Firth, *Acute Sensitivity of Selected Australian Freshwater Animals to Copper and Zinc,* Australian Water Resources Council, Technical Paper No. 81, Australian Government Publishing Service, Canberra, 1983.

49. G. C. Gerloff and W. H. Gabelmann, in *Inorganic Plant Nutrition,* Vol. 15B (A. Läuchli and R. L. Bieleski, eds.), Springer, Berlin, 1983, p. 453 ff.

50. R. C. Schenk, *Mar. Biol. Lett., 5,* 19 (1984).

51. M. R. Reeve, G. D. Grice, Y. R. Gibson, M. A. Walter, K. Darcy, and T. Ikeda, in *Effect of Pollutants on Aquatic Organisms* (A. P. M. Lockwood, ed.), Cambridge University Press, Cambridge, 1976, p. 145 ff.

52. A. Fogels and J. B. Sprague, *Water Res., 11,* 811 (1977).

53. M. Waldechuck, in *Pollution and Physiology of Marine Organisms* (F. J. Vernberg and W. B. Vernberg, eds.), Academic Press, New York, 1974, p. 1 ff.

54. R. Gächter, *Schweiz. Z. Hydrol., 38,* 97 (1976).

55. P. Bossard and R. Gächter, *Schweiz. Z. Hydrol., 41,* 261 (1979).

56. J. J. Rosko and J. W. Rachlin, *Bull. Torrey Bot. Club, 102,* 100 (1975).

57. A. J. Horne and C. R. Goldmann, *Science, 183,* 409 (1974).

58. J. J. Goering, D. Boisseau, and A. Hattori, *Bull. Mar. Sci., 27,* 58 (1977).

59. N. M. Morel, J. G. Rueter, and F. M. Morel, *J. Phycol., 11,* 43 (1978).

60. J. G. Rueter, Jr., and F. M. M. Morel, *Limnol. Oceanogr., 26,* 67 (1981).

61. C. A. Price and J. W. Quigley, *Soil Sci., 101,* 11 (1966).

62. B. R. Lumsden and T. M. Florence, *Environ. Technol. Lett., 4,* 271 (1983).

63. M. R. Reeve, J. C. Gamble, and M. A. Walter, *Bull. Mar. Sci., 27,* 92 (1976).

64. J. Urech, *Schweiz. Z. Hydrol., 41,* 247 (1979).

65. W. K. Besch, M. Ricard, and R. Cantin, *Int. Rev. Ges. Hydrobiol., 57,* 39 (1979).

66. T. J. Hara, S. B. Brown, and R. E. Evans, in *Aquatic Toxicology* (J. O. Nriagu, ed.), John Wiley, New York, 1983, p. 247 ff.

67. H. Babich and C. Stotzky, *Crit. Rev. Microbiol., 8,* 99 (1980/ 1981).

68. P. M. Mehrle, F. L. Mayer, and W. W. Johnson, in *Aquatic Toxicology and Hazard Evaluation* (F. L. Mayer and J. L. Hamlink, eds.), ASTM, Philadelphia, 1977, p. 269 ff.

69. S. L. Sosnowski, D. J. Germond, and J. H. Gentile, *Water Res., 13,* 449 (1979).

70. J. Cairns, Jr., A. L. Buikema, Jr., A. G. Heath, and B. C. Parker, *Va. Water Resour. Res. Cent. Bull., 106,* 88 (1978).

71. W. K. Seim, L. R. Curtis, S. W. Glenn, and G. A. Chapmann, *Can. J. Fish. Aquat. Sci., 41,* 433 (1984).

72. R. Gächter and A. Mares, *Schweiz. Z. Hydrol., 41,* 228 (1979).

73. W. H. Thomas and D. L. R. Seibert, *Bull. Mar. Sci., 27,* 23 (1977).

74. J. I. Mangi and G. J. Schumacher, *Midl. Nat., 102,* 134 (1979).

75. J. S. Weis and P. Weis, in *Aquatic Toxicology* (J. O. Nriagu, ed.), John Wiley, New York, 1983, p. 189 ff.

76. E. E. Kenaga and C. A. I. Goring, in *Aquatic Toxicology and Hazard Evaluation* (J. G. Eaton, P. R. Parrish, and A. C. Hendricks, eds.), ASTM, Philadelphia, 1980, p. 24 ff.

77. J. L. Hamelink, in *Aquatic Toxicology and Hazard Evaluation* (F. L. Mayer and J. L. Hamelink, eds.), ASTM, Philadelphia, 1974, p. 24 ff.

78. D. A. Wright, *Appl. Biol.*, *3*, 331 (1978).

79. D. Saward, A. Stirling, and G. Topping, *Mar. Biol.*, *29*, 351 (1975).

80. M. Aubert, R. Bittel, F. Laumond, M. Romeo, D. Donnier, and M. Barelli, *Rev. Int. Oceanogr. Med.*, *33*, 7 (1974).

81. B. Wachs, *Münchner Beiträge zur Abwasser-, Fischerei- und Flussbiologie,* Vol. 32, Oldenburg, Munich, 1980, p. 387 ff.

82. R. Gächter and W. Geiger, *Schweiz. Z. Hydrol.*, *41*, 277 (1979).

83. W. H. Thomas, O. Holm-Hansen, D. L. R. Seibert, F. Azam, R. Hodson, and M. Takahashi, *Bull. Mar. Sci.*, *27*, 34 (1977).

84. J. R. Beers, G. L. Steward, and K. D. Hoskins, *Bull. Mar. Sci.*, *27*, 66 (1977).

85. R. F. Vaccaro, F. Azam, and R. E. Hodson, *Bull. Mar. Sci.*, *27*, 17 (1977).

86. E. Eichenberger, F. Schlatter, H. Weilenmann, and K. Wuhrmann, *Verh. Int. Ver. Limnol.*, *21*, 1131 (1981).

87. R. Gächter and J. Urech, in *Trace Element Speciation in Surface Water and Its Ecological Implications* (G. G. Leppard, ed.), Plenum Press, New York, 1983, p. 137 ff.

88. H. R. Bürgi, *Schweiz. Z. Hydrol.*, *36*, 1 (1974).

89. E. Eichenberger, unpublished observations.

90. A. Sode, *Arch. Hydrobiol. Suppl.*, *65*, 134 (1983).

4

Metal Ion Speciation and Toxicity in Aquatic Systems

Gordon K. Pagenkopf
Department of Chemistry
Montana State University
Bozeman, Montana   59717

1. INTRODUCTION

The observations that elevated trace metal concentrations in streams, rivers, and lakes resulted in large quantities of dead fish has initiated many laboratory toxicity studies [1]. Initial laboratory conditions have been improved and correspondingly more reliable results are available. Some of the parameters that have been investigated include temperature [2], pH [3,4], water hardness [5-9], fish type [10], and trace metal complexation [8,11-18]. As a consequence, several models have been put forth to account for variations

in the amounts of metal needed to give rise to specified toxicity
[12,18-20].

Recent studies [8,21,22] indicate that trace metal speciation
in field samples is very different from that observed in laboratory
samples. The consequence is that a much more detailed chemical
composition of a sample must be obtained before accurate chemical
speciation can be determined. In addition, major changes in metal
toxicity will be dependent upon speciation.

This chapter utilizes the results of many studies involving
the toxicity of copper, cadmium, lead, and zinc to fishes. Analyses
of the literature identifies at least four general conclusions:
(1) Some chemical species of a metal are more toxic than others,
(2) an increase in water hardness tends to reduce the metal toxicity,
(3) some metals are more toxic than others, and (4) naturally occurring materials often reduce the concentrations of toxic species.
These parameters are not the only ones affecting toxicity, but they
can be utilized to rationalize many observed metal toxicity results.
In general, a model based on the first three items [18] is modified
to include the interaction of toxic trace metals with naturally
occurring complexing and adsorbing material.

2. CHEMICAL MODEL

A chemical model has been developed to interpret metal toxicities
observed at various hardness, alkalinity, and pH values [18]. This
model is applicable to laboratory-type studies that utilize waters
having minimal component variation. Specifically the presence of
complexing material in natural waters can make dramatic trace metal
speciation changes and correspondingly apparent fish toxicity changes.
The physiological results associated with short-term trace metal toxicity are not the main function of this work; however, extensive variations in exposure concentrations can cause a large variation in
mucous secretion [23-25]. Trace metals may concentrate in gill

surfaces [26,27]. This model focuses on trace metal speciation and subsequent fish toxicity.

The toxic trace metals are usually divalent, for example, Cu^{2+}, Zn^{2+}, Cd^{2+}, and Pb^{2+}, and thus these will be referred to as TM^{2+}. Specific species will be identified when the metals are discussed.

Gill surfaces contain phospholipids that can provide coordination sites for the trace metal species. This approach permits the application of conventional parameters as shown here in Eq. (1):

$$TM^{2+} + \equiv S^{m-} \xrightleftharpoons{K_m} \equiv S \cdot TM^{-m+2} \tag{1}$$

In this particular case the gill surface is a nondefined Lewis base and the cation is a Lewis acid. Additional metal species will be included subsequently. In this case $\equiv S^{m-}$ designates the surface and $\equiv S \cdot TM^{-m+2}$ designates a surface complex. The reactions for the four metals are generally rapid and thus an equilibrium expression can be written:

$$K_{TM} = \{\equiv S \cdot TM^{-m+2}\}/([TM^{2+}]\{\equiv S^{m-}\}) \tag{2}$$

Rearrangement of Eq. (2) provides the following:

$$\{\equiv S \cdot TM^{-m+2}\} = K_{TM}[TM^{2+}]\{\equiv S^{m-}\} \tag{3}$$

which indicates that the activity of the complexed metal is proportional to the free concentration and the size of the gill surface. The brackets designate a concentration in moles per liter, whereas the braces designate moles per kilogram. With $\equiv S \cdot TM^{-m+2}$ being a toxic configuration, Eq. (3) indicates a linear relationship between toxicity and the metal species concentration. If a sizable fraction of the $\equiv S^{m-}$ sites are complexed, a nonlinear response and a short exposure time would be observed. In general, most 96-hr toxicity studies exhibit linear relationships between mortality and toxic species concentrations.

The most critical portion of the model is an assessment of the trace metal toxic species. In many cases a large fraction of the

trace metal is complexed by a ligand other than water and correspondingly forms a species that is not toxic or only slightly toxic.
In general, the free metal species, TM^{2+}, and the hydroxide complexes,
$TM(OH)^+$ and $TM(OH)_2 \cdot aq$, tend to be the most toxic, whereas a complex
with carbonate, $TM(CO_3) \cdot aq$, exhibits minimal toxicity. This was
observed in laboratory studies and will be critical in natural water
systems. However, the presence of natural complexing ligands will
permit formation of many complexes that will also tend to reduce
the metal toxicity.

Previously only one chemical reaction has been considered,
Eq. (1), and as a consequence the chemical speciation was much less
than what occurs in natural waters. The fraction that one metal
species, say, TM^{2+}, makes of the total is defined as

$$\alpha_{TM} = \frac{[TM^{2+}]}{[TM_T]} \tag{4}$$

Combining this with Eq. (3) yields

$$\{\equiv S \cdot TM^{-m+2}\} = K_{TM} \alpha_{TM} [TM_T] \{\equiv S^{m-}\} \tag{5}$$

which takes into account that many species can be associated with
the gill surface. The interactions can be adsorption and/or absorption. A generalization of Eq. (5) provides the following:

$$\{\equiv S \cdot TM_{(i)}\} = K_{TM_{(i)}} \alpha_{TM_{(i)}} [TM_T] \{\equiv S^{m-}\} \tag{6}$$

For some trace metals there are many different species in solution,
for example, copper, whereas others, such as cadmium, have fewer
species in natural water systems. Specifics for each metal will be
presented later. A general equation can be written with the following format, however:

$$TM_T = TM^{2+} + TM(OH)^+ + TM(OH)_2 \cdot aq + TM(CO_3) \cdot aq$$

$$+ \Sigma TM\text{-}org_{(i)} + \Sigma TM\text{-}surface_{(i)} \tag{7}$$

Laboratory studies generally have minimal or at least controlled contribution from the last two terms. This is not the case for many natural water systems.

Trace metal toxicity tends to decrease when the water contains higher concentrations of the hardness metals, namely, calcium and magnesium. The concentration range of these metals is generally from 10^{-4} to 10^{-2} M. In addition, the complexation stability constants for these metals are similar. This leads to the following general relationship:

$$\equiv S_T = \equiv S^{m-} + \equiv S \cdot M^{-m+2} + \equiv S \cdot H^{-m+1} + \equiv S \cdot TM^{-m+2} \qquad (8)$$

In this case M^{2+} represents $Ca^{2+} + Mg^{2+}$. It is assumed that protonation occurs below pH 6, and the number of sites occupied by the trace metal species is small compared to those free and those complexed by the hardness metals. Thus

$$\equiv S_T = \equiv S^{m-} + \equiv S \cdot M^{-m+2} \qquad (9)$$

and the equilibrium relationship for generalized calcium and magnesium complexation is given by

$$K_M = \{\equiv SM^{-m+2}\}/([M^{2+}]\{\equiv S^{m-}\}) \qquad (10)$$

Combining Eqs. (9) and (10) yields the following:

$$\equiv S^{m-}/\equiv S_T = 1/(1 + K_M[M^{2+}]) = CIF \qquad (11)$$

where the ratio $\equiv S^{m-}/\equiv S_T$ is defined as the competitive interaction factor (CIF). Combining Eqs. (6) and (11) provides

$$ETC = \frac{\equiv S \cdot TM_{(i)}}{(K_{TM_{(i)}})(\equiv S_T)} = \frac{[TM_T] \cdot \alpha_{TM_{(i)}}}{1 + K_M[M^{2+}]} = CIF \cdot \Sigma[\text{toxic species}]_{(i)} \qquad (12)$$

This expression combines the appropriate toxic metal species and the hardness protection to generate the effective toxicant concentration (ETC). This model is subsequently applied to copper, cadmium, lead, and zinc.

3. APPLICATION TO COPPER TOXICITY

There have been a large number of copper toxicity studies, most of
which were laboratory controlled. In many cases there was minimal
control of the chemical composition other than total copper present
in static or flowing water. There are three studies utilizing rainbow trout [5,7,28] and one utilizing cutthroat trout [4] that provided good control of the chemical parameters. In general, the
studies have focused on a pH change, from 6 to 9, and a water hardness variation, from 10 to 371 mg/liter, as $CaCO_3$. These results
are summarized in Table 1. Of specific interest is the total copper
present and that present as the apparent toxic species, Cu^{2+}, $CuOH^+$,
and $Cu(OH)_2 \cdot aq$.

Analysis of the data presented in Table 1 indicates that there
is a tremendous variation in the total copper present. For example,
18-516 µg/liter for rainbow trout and 16-367 µg/liter for the cutthroat trout. In addition, there is a sizable pH and bicarbonate
concentration variation. All three of these parameters determine
the effective toxicant concentration. A critical factor is the
formation of complexes that are not toxic, or, if they are, their
toxicity is much less than that of the three species used to calculate the ETC. The other factor is the competition between the toxic
species and the water hardness cations Ca^{2+} and Mg^{2+}. This factor
is very significant and can be greater than 10 in extremely hard
water.

The laboratory studies are in fact extremely well controlled
when compared to the chemical variability that can and does occur
in natural water systems. Some of the important components of the
natural systems are humic and fulvic acids. In addition, the presence of suspended material, whether organic or inorganic, can cause
a dramatic change in copper speciation and distribution.

A good example of how copper can be distributed between the
solution phase and the adsorbed phase has been provided by Mouvet
and Bourg [8]. Their study involved a river that exhibited sizable

TABLE 1

Total Copper and Predicted ETC Values

Hardness/alkalinity as CaCO$_3$ (mg/liter)	pH	Cu$_T$ (µg/liter)	ETC (µg/liter)	Reference
Rainbow trout				
72/10	7.1	19	7.87	7
99/10	7.0	54	6.52	7
49/28	7.3	48	6.10	7
98/28	7.2	78	5.18	7
12/51	7.4	18	2.22	7
97/51	7.3	96	3.69	7
300/205	7.3	96	3.69	20
32/—	6.0	22.4	9.77	5
101/—	6.0	40.0	6.72	5
371/—	6.0	82.2	6.58	5
101/—	7.0	47.9	5.50	5
361/—	7.0	298	8.22	5
31/—	8.0	30.0	7.13	5
317/—	8.0	516	6.93	5
360/—	8.0	309	4.16	5
30/—	9.0	30.0	6.51	5
98/—	9.0	85.9	5.98	5
		average = 6.06 µg/liter		
Cutthroat trout				
205/178	7.73	367.	1.65	4
205/77.9	7.61	186.	2.44	4
160/26.0	7.53	36.8	3.04	4
70/174	8.54	232.	2.20	4
70/70	7.40	162.	5.54	4
74/23	7.57	73.6	2.98	4
18/183	8.07	91.0	0.81	4
18/78	8.32	44.4	3.75	4
26/20	7.64	15.7	2.24	4
		average = 2.72 µg/liter		

TABLE 2

Distribution of Copper Between Adsorbed and Solution Species [8]

Species	Percent	Species	Percent
Cu^{2+}	0.3	$CuSO_4 \cdot aq$	<0.1
$Cu(OH)^+$	0.1	CuHA	<0.1
$Cu(OH)_2 \cdot aq$	6.4	Cu^{2+} (adsorbed)	87
$CuCO_3 \cdot aq$	4.4	Cu-HA (adsorbed)	1.5
$CuHCO_3^+$	0.1		

variations in suspended matter, from 1 to 843 mg/liter. The percentage distribution of various chemical species was probably similar.
A detailed distribution of one of the samples is shown in Table 2. A
sum of the concentrations of the three most toxic species, Cu^{2+}, $CuOH^+$,
and $Cu(OH)_2 \cdot aq$, provides 1.09 µg/liter, which, when adjusted for competition from Ca^{2+} and Mg^{2+}, would be about a factor of 40 below the
average ETC shown in Table 1. If the copper were not adsorbed, the
toxic species would be about a factor of 5 less than the ETC value
and would probably show some toxicity.

There are many waters receiving copper pollution that do not
have as much suspended material or complexing agents. As a consequence the toxic copper species will constitute a larger fraction
of the total copper. This can be demonstrated by looking at how
copper is distributed between the toxic and nontoxic species. There
are many suspended solid-phase materials that can adsorb copper.
Some of these are hydroxides of Fe(III), Al(III), and silica. In
addition, the solution phase may include humic and fulvic acids [17]
and other ligands [16] that are much stronger chelators than some
that have been identified [29].

A series of calculations have been developed to predict the
copper toxicity to fish as the adsorption and complexation capacities
vary. The conditions which are listed in Table 3 are utilized to
predict the concentrations of the toxic copper species. The concen-

TABLE 3

Conditions Utilized to Calculate Copper Species Distribution

$T = 25.0°C$	$HA;\ 10^{-6}\ M$
$pH = 8.0$	Suspended matter; $1 \cdot 10^{-5}\ M$
$[Ca^{2+}] + [Mg^{2+}] = 1.0 \cdot 10^{-3}\ M$	$\log K_{CuOH} = 6.3;\ \log \beta_2\ [Cu(OH)_2 \cdot aq] = 12.6$
$[HCO_3^-] = 2.0 \cdot 10^{-3}\ M$	$\log K_{CuCO_3} = 6.77;\ \log K_{Cu-HA} = 5.0$
$Cu_T;\ 10^{-8} - 10^{-5}\ M$	$\log K_{Cu-SM} = 6.2;\ \log K_{CaCO_3} = 3.2$

tration sums of the toxic species and other species are plotted as a function of the total copper in the system (see Fig. 1).

Comparison of the plots in Fig. 1 indicates that the toxic copper species constitute the smallest fraction of the total copper. In general, the two complexes $CuCO_3 \cdot (aq)$ and Cu-HA constitute about 47%, that associated with the solid phase accounts for 33%, and the remainder, which is the sum of the toxic species present, approximately 20%. The distribution of copper between the three general groups may occur fairly rapidly. A decrease in the concentrations of the complexing ligands and the adsorbing surface will in general raise the amount of toxic material present, provided that the total copper is kept constant. If the pH goes down, the stability of $CuCO_3 \cdot aq$ and other complexes will decrease. If, for example, the total copper concentration remains the same and the pH drops from 8.0 to 7.0, the carbonate complex concentration will drop by a factor of 10. Some of the released copper will be adsorbed, however; the rest can be present as toxic species and thus some short-term toxicity could be inflicted. A 10-fold increase in effective stability constants for suspended material will greatly shift the equilibrium distribution. In this case the fraction present as toxic species will decrease and correspondingly the fish will be

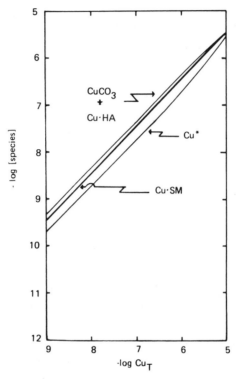

FIG. 1. Distribution of copper between toxic and nontoxic components
$Cu^* = Cu^{2+} + Cu(OH)^+ + Cu(OH)_2 \cdot aq$, Cu$\cdot$SM = copper suspended matter;
Cu^* = toxic copper species).

able to survive in waters containing much more copper. We have
recently made an observation where virtually all of the copper was
present in the suspended material.

In natural systems there is the possibility of having many
different types of material present in the suspended phase. Some
of these will be better-complexing (adsorbing) units and thus will
provide more protection.

4. APPLICATION TO CADMIUM TOXICITY

The chemistry of cadmium(II) in water exhibits some factors that are similar to copper(II) and other metals, but there are some differences that are critical to fish toxicity. As with copper(II), complexation and adsorption of cadmium will play a major role in determining the effective toxicity. A stability summary of cadmium and other metal complexes with a variety of ligands is listed in Table 4. In general, the cadmium complexes are not as stable as those of copper and lead. This will have a major effect on the distribution of cadmium between the toxic and nontoxic complexes. Since the stability constants for the soluble species are in the 10^4 range, a sizable quantity of this material has to be present before there is a major shift in the species distribution. A classic example is a comparison of cadmium and copper distributions in carbonate complexes. For example, if the carbonate concentration were 10^{-5} M the ratio of

TABLE 4

Representative Stability Constants Given in Logarithms

Ligand	Cu(II)	Cd(II)	Pb(II)	Zn(II)	References
NH_3; K_1	4.25	2.51	—	2.18	31
Glycine; K_1	8.62	4.8	5.47	5.52	31
Oxalate; K_1	6.19	4.0	4.0	4.68	31
OH^-; K_1	6.3	4.6	6.2	4.4	30
OH^-; β_2	12.6	8.9	10.9	10.0	30
Carbonate	6.77	4.0	6.2	4.8	30
HA; pH = 8	5.0	5.0	5.3-8	5.3	8, 33
Surface species	-1.8	-3.7	-1.7	-3.6	8

$CuCO_3 \cdot aq$ to Cu(II) would be 59, whereas the ratio of $CdCO_3 \cdot aq$ to
Cd(II) would be 0.1. The difference indicates a sizable variation
in the speciation of these two metals under the same conditions.
The same general behavior is predicted when other stability constants are compared. The stability constant for the ligands,
ammonia, glycine, and oxalate are considerably less for Cd(II)
than for Cu(II) (Table 4).

Interaction of Cd with surface material occurs; however, it
does not appear to have the stability that would permit a large
fraction of the total cadmium present to be nontoxic. A conditional
stability constant [8] for adsorption is $10^{-3.7}$ at pH 8.0. This
would provide an effective stability constant of $10^{4.3}$. At a 10^{-5} M
surface-active species concentration, which is probably too high, an
increase of approximately 20% in total cadmium could be expected.
This is small and thus more thermodynamically stable species need
to be formed before a sizable fraction of the cadmium can be present
as nontoxic soluble species or there must be sizable association of
cadmium(II) with the suspended solid phase. For example, the 96-hr
ETC value for rainbow trout is 56 µg/liter [32]. If the surface
complexing material concentration were 10^{-5} M, an approximate 50%
total metal increase, from 56 to 72 µg/liter, could be tolerated
[18]. This is not a very large increase and thus the toxicity of
cadmium is not as sensitive to the chemistry of the water. Calculations do indicate that chemical stability increased by approximately a factor of 10 would make a sizable change in the speciation
and correspondingly reduce the potential toxicity. This probably
occurs in many surface water systems.

5. APPLICATION TO LEAD TOXICITY

The inorganic chemistry of lead in natural water is somewhat more
pronounced than that observed for cadmium, but it is not quite as
detailed as that for copper. Very high lead concentrations would
make a significant difference, however. The hydroxide species that

form in the near neutral pH range, $PbOH^+$ and $Pb(OH)_2 \cdot aq$, will be
fairly significant contributors to the other toxic species, Pb^{2+}.
The stability constants [31] for these two species are $K_1 = 10^{6.2}$
and $\beta_2 = 10^{10.9}$. As a consequence, the ratio of $Pb(OH)^+$ to Pb^{2+}
will be unity when the pH is equal to 6.8 and it will increase to
a factor somewhat greater than 10 if the pH is near 8. If the pH
of the natural water system is around 7, the ratio of $Pb(OH)_2 \cdot aq$
to Pb^{2+} will be too small for the detection of any toxic contribution from $Pb(OH)_2 \cdot aq$. If the pH of the solution is in the 8.5
region, there will be approximately equal concentrations of Pb^{2+}
and $Pb(OH)_2 \cdot aq$, however.

   Another ligand that will make a significant contribution to
the soluble species is carbonate. The stability constant is $10^{6.3}$
[30], which is comparable to the value observed for copper. If the
total inorganic carbon concentration is $2 \cdot 10^{-3}$ at pH 8.0, then the
ratio $PbCO_3 \cdot aq/Pb^{-+}$ is predicted to be approximately 16. The ratio
of these two species and correspondingly the carbonate complexation
capacity decrease as the solutions become more acidic. For example,
at pH 7 the ratio of these two species would have decreased to a
value near 2.

   For fish to tolerate elevated concentrations of lead in a
natural water system there have to be other effective complexing
agents. Humic and fulvic acids can be contributors to this group,
since the stability of the lead complexes seems to be fairly large.
The stability constants for the interaction of lead with humic acid
exhibit considerable variation, however, for example, values of
$10^{5.3}$ at pH 8 [8] up to a value of $10^{8.0}$ [33]. If there is 1 mg/
liter humic acid (HA) in the water with an effective gram formula
weight of 1000, a sizable amount of lead could be complexed. With
these conditions the HA concentration would be 10^{-6} M, and if the
10^8 stability constant were applicable, essentially all of the humic
acid would complex Pb^{2+}. This would raise the total nontoxic species
concentration and in fact could cause a shift from apparent toxicity
to nontoxicity. It is likely, however, that the humic acid-lead
stability constants are less than the 10^8 value, that is, pH $\cdot$ 0.625

[8], and thus the degree of lead complexation by humic acid will decrease proportionality. In fact, fresh waters may not contain, enough humic acid to cause a very significant complexation contribution.

The predicted conditional stability constant for the interaction of lead with suspended solid oxides is $10^{-1.7}/[H^+]$ [8], which at pH 8 provides an effective stability constant of $10^{6.3}$. A suspended adsorption material concentration of 10^{-5} M would produce a significant ratio $Pb^{2+}_{(adsorbed)}/Pb^{2+}$ of about 20. This will, of course, permit an increase in the total amount of lead that can be in the solution phase.

The effective toxic concentration for lead lies in the range of 4 µg/liter [18,34]. Of this Pb^{2+} and $Pb(OH)^+$ will be the major species, since the inherent solubility of $Pb(OH)_2 \cdot aq$ will not exceed approximately $10^{-9.9}$ M. The presence of humic acid, carbonate, and adsorption material have the capability of dramatically raising the total concentration of lead present but still maintaining a concentration of toxic species below the 96-hr LC_{50}.

6. APPLICATION TO ZINC TOXICITY

The speciation of zinc(II) in natural waters, pH 6.5-8.5, does not include a major contribution from the hydroxide species. For example, the ratio of $Zn(OH)^+$ to Zn^{2+} at pH 8.0 is 0.025, whereas the ratio of $Zn(OH)_2 \cdot aq$ to Zn^{2+} is 0.01. As a consequence, these two hydroxide species will make a minimal contribution to the fish toxicity. The other inorganic complexing ligand, CO_3^{2-}, will be able to complex a fraction of the total zinc. For example, if the bicarbonate concentration is $2.0 \cdot 10^{-3}$ M at pH 8.0, one predicts a ratio of $ZnCO_3 \cdot aq$ to Zn^{2+} of 0.63. This ratio is much less than that observed for copper and lead and thus the degree of complexation is also much less. Humic acid is also capable of complexing zinc. In this case a 10^{-6} M humic acid solution would generate a $Zn-HA/Zn^{2+}$ ratio of 0.2. Based on these complexation reactions, it is indicated that over 80% of the soluble zinc would be present as the toxic species Zn^{2+}. If, however, the humic acid concentration were to increase to 10^{-5} M or if the stability constant were to increase by a factor of 10, a sizable change in the species distribution would occur. In this case the ratio of complexed to noncomplexed zinc would be 2.

The interaction of zinc with surface material [8] has a conditional constant of $10^{-3.6}$, which at pH 8 would equal $10^{4.4}$. This value is considerably less than that observed for Cu^{2+}, but it can still make a sizable contribution to the distribution of the total zinc. If the suspended material is present in the 10^{-5} M range, then approximately 25% of the zinc could be present in this material. Combination of the complexed and adsorbed zinc indicates that between 25 and 50% of the total zinc could be present in nontoxic forms. As a consequence, total zinc in the range 250-380 μg/liter could be observed before the 96-hr LC_{50} for rainbow trout was exceeded. That value is 190 μg/liter [35].

There have been some recent observations that zinc is readily precipitated from natural waters [21], and thus it is conceivable that the speciation will occur but the total zinc available will be significantly reduced.

7. CONCLUSION

A majority of the trace metal toxicity studies have been conducted in the laboratory environment, where the parameters can be controlled. Application of these results to field observations can be fairly straightforward in some cases; however, it is difficult in most cases. Of major concern in the field is variation of the pH and the presence of low concentrations of material that can complex and adsorb the toxic metal species. Analytical techniques are available to obtain good estimates of the trace metal speciation; however, they require many experiments that are generally not done. As a consequence, the total trace metal concentration in a field water sample implies little regarding the metals' toxicity. The presence of adsorbing and complexing material in natural water makes the correlation between total metal and toxic species very difficult. As a consequence, the field studies must be more detailed before the metal toxicity potential can be assessed.

ABBREVIATIONS AND DEFINITIONS

CIF	competitive interaction factor
ETC	effective toxicant concentration
HA	humic acid
LC_{50}	concentration of toxic metal leading to the death of 50% of a population
SM	suspended matter
$\equiv S^{m-}$	surface with Lewis base properties
TM^{2+}	divalent toxic trace metal ion
[ ]	designate mole per liter
{ }	designate mole per kilogram

REFERENCES

1. P. Doudoroff and M. Katz, *Sewage Ind. Wastes, 25*, 802 (1953).

2. P. V. Hobson and J. B. Sprague, *J. Fish Res. Board Can., 32*, 1-10 (1975).

3. D. I. Mount, *Air Water Pollut. Int. J., 10*, 49-56 (1966).

4. C. Chakoumakos, R. C. Russo, and R. V. Thurston, *Environ. Sci. Technol., 13*, 213-219 (1979).

5. R. S. Howarth and J. B. Sprague, *Water Res., 12*, 455-462 (1978).

6. D. H. Stendahl and J. B. Sprague, *Water Res., 16*, 1479-1488 (1982).

7. T. G. Miller and W. C. Mackay, *Water Res., 14*, 129-133 (1980).

8. C. Mouvet and A. C. M. Bourg, *Water Res., 17*, 641-649 (1983).

9. D. I. Mount and C. E. Stephan, *J. Fish. Res. Board Can., 26*, 2449-2457 (1969).

10. P. D. Anderson and L. J. Weber, *Toxic Appl. Pharmacol.*, *33*, 471-483 (1975).

11. R. Lloyd and D. W. M. Herbert, *Inst. Public Health Eng. J.*, *61*, 132-145 (1962).

12. G. K. Pagenkopf, R. C. Russo, and R. V. Thurston, *J. Fish. Res. Board Can.*, *31*, 462-465 (1974).

13. T. L. Shaw and V. M. Brown, *Water Res.*, *8*, 37-382 (1974).

14. V. Zitko, W. V. Carson and W. G. Carson, *Bull. Envir. Contamin. Toxicol.*, *10*, 265-271 (1973).

15. R. W. Andrew, K. E. Biesinger, and G. E. Glass, *Water Res.*, *11*, 309-315 (1977).

16. C. M. G. van den Berg, P. T. S. Wong, and Y. K. Chau, *J. Fish. Res. Board Can.*, *36*, 901-905 (1979).

17. W. Fish and F. M. M. Morel, *Can. J. Fish. Aquat. Sci.*, *40*, 1270-1277 (1983).

18. G. K. Pagenkopf, *Environ. Sci. Technol.*, *17*, 342-347 (1983).

19. P. V. Hodson, J. W. Hilton, B. R. Blunt, and S. J. Slinger, *Can. J. Fish. Aquat. Sci.*, *37*, 170-176 (1980).

20. D. P. H. Laxen and R. M. Harrison, *Water Res.*, *15*, 1053-1065 (1981).

21. G. E. Niembe and G. F. Lee, *Water Res.*, *16*, 1373-1378 (1982).

22. C. Houba, J. Remacle, D. Dubois, and J. Thorez, *Water Res.*, *10*, 1281-1286 (1983).

23. R. Eisler and G. R. Gardner, *J. Fish Biol.*, *5*, 131-142 (1973).

24. R. Eisler, *J. Fish Biol.*, *6*, 601-602 (1974).

25. J. F. Skidmore and P. W. A. Tovell, *Water Res.*, *6*, 217-230 (1972).

26. G. R. Phillips and R. C. Russo, *Metal Bioaccumulation in Fishes and Aquatic Invertebrates: A Literature Review*, EPA-600/3-78-103, U.S. Environmental Protection Agency, Duluth, Minnesota (1978).

27. W. A. Brungs, E. N. Leonard, and J. M. McKim, *J. Fish. Res. Board Can.*, *30*, 583-586 (1973).

28. D. Calamari and R. Marchetti, *Water Res.*, *7*, 1453-1464 (1973).

29. J. Buffle, *Anal. Chim. Acta*, *118*, 29-44 (1980).

30. R. M. Smith and A. E. Martell, *Critical Stability Constants*, Vol. 4, Plenum Press, New York, 1976.

31. L. G. Sillén and A. E. Martell, *Stability Constants of Metal-Ion Complexes*, Special Publication No. 25, The Chemical Society, London, 1971.

32. D. Calamari, R. Marchetti, and G. Vailatis, *Water Res.*, *14*, 1421-1426 (1980).

33. F. J. Stevenson, *Soil Sci. Soc. Am. J.*, *40*, 665-672 (1976).

34. P. H. Davies, J. P. Goettl, Jr., J. R. Sinley, and N. F. Smith, *Water Res.*, *10*, 199-206 (1976).

35. G. W. Holcombe and R. W. Andrew, EPA 600/3-78-094, U.S. Environmental Protection Agency, Washington, D.C., 1978.

5

Metal Toxicity to Agricultural Crops

Frank T. Bingham, Frank J. Peryea,* and Wesley M. Jarrell
Department of Soil and Environmental Sciences
University of California
Riverside, California 92521

*Present affiliation: Tree Fruit Research Center, Washington State
University, Wenatchee, Washington 99163

1. INTRODUCTION

Trace metals in the geochemical environment may all be toxic if
present at elevated bioavailable concentrations. A number of these
metals are essential for the growth of plants. Obviously, in a
chapter of the length permitted we cannot address all trace metals.
We have therefore elected to discuss five metals—three of which are
essential for the growth of crops—namely, copper, manganese, and zinc,
and two that are not considered to be essential, cadmium and nickel.

Copper, manganese, nickel, and zinc toxicities to crops occur
under natural conditions in the form of reduced yields. Although
phytotoxic effects of Cd have been frequently observed in greenhouse
studies, no documented case exists where Cd in soil has reduced the
yields of field-grown commercial crops. Phytotoxicities commonly occur
on crops grown on acid soils and are rare on neutral and alkaline soils.

Industrial uses of the metals are many and widespread. Therefore phytotoxicities and excessive accumulation in soils are commonly
observed near point sources. Much of the information on effects of
trace metals in soils on crop growth and composition has been accumulated from soils contaminated by point sources, particularly mining
and smelting operations, as well as the disposal or recycling of
municipal sewage sludges on land.

The following discussion centers on factors influencing the
concentration of available metals in soils for a wide variety of
crop species. Metal uptake and accumulation by plants are dependent

not only upon the concentration of available metals in soil but also
upon the plant species. Phytotoxic concentrations in soils and
plants derived from greenhouse experiments are also discussed.

2. CADMIUM

2.1. Introduction

Cadmium (Cd) is a trace metal that is toxic to both animals and
plants when present at excessive bioavailable concentrations in the
environment. Cadmium also has a tendency to persist in the soil.
Although there are no observations of Cd toxicity (reduced crop
yields) to crop species in the field, Cd is accumulated in plants.
Thus factors influencing Cd availability to food crops are under
intensive study because of the potential hazard to humans consuming
plant food products containing elevated levels of Cd.

2.2. Forms in Soil

The Cd content of unpolluted soil is closely related to that of the
parent material. The Cd content of the earth's crust averages 0.15
mg/kg. Cadmium concentrations in igneous, metamorphic, and sedimen-
tary rocks range from 0.001 to 1.8, 0.04 to 1.0, and 0.1 to 11.0
mg/kg, respectively. Soils derived from such parent materials contain 0.1-0.3, 0.1-1.0, and 3.0-11.0 mg/kg cadmium [1,2]. Recently
Holmgren et al. [3] summarized the Cd content of 3202 surface soils
(excluding organic soils) collected from the major crop-producing
regions in the United States. They found soil Cd to range from
0.03 to 0.9 mg/kg, with an average of 0.27 mg/kg.

The species in soil solutions of acid and calcareous soils,
respectively, are Cd^{2+}, $CdCl^+$, and $CdSO_4^0$; and Cd^{2+}, $CdCl^+$, $CdSO_4^0$,
and $CdHCO_3^+$. Cadmium does not complex with organic ligands to any
extent in cultivated soils [4-6]. The adsorbed Cd of Cd-treated

soils varies from 90 to 99% of the Cd added at relatively low rates
(<20 mg Cd per kg soil) [7].

2.3. Availability

2.3.1. *Substrate Concentration and Plant Species*

Response to substrate Cd is highly dependent upon the plant species
as well as the concentration of available Cd. Page et al. [8] reported the effect of Cd levels in nutrient solutions on Cd uptake
and growth for a variety of common crop species. Test plants were
established in aerated solution culture tanks containing 100 liters
of a complete nutrient solution. Cadmium, as $CdSO_4$, was then added
to the solutions, producing Cd levels ranging from 0.1 to 10.0 mg/
liter. The plants were grown for 3 weeks in the presence of Cd,
harvested, and analyzed for leaf Cd content. Table 1 summarizes the
results. The data demonstrate a strong plant species effect with
respect to threshold concentrations of Cd in solution and uptake as
judged by leaf Cd levels. For example, concentrations ≤ 0.10 mg/liter
reduced growth 25% for beet (*Beta vulgaris* L.), field bean (*Phaseolus
vulgaris* L.), and turnip (*Brassica rapa* L.). In marked contrast, a
level of 4.00 mg Cd per liter was required for similar growth reduction in cabbage (*Brassica oleracea* L.). These data also show that
no general relationship exists between leaf Cd content and tolerance
to Cd.

Bingham et al. [9] grew a number of agricultural crops in
potted soils using a pH 7.5 sandy loam treated with municipal sewage
sludge enriched with various amounts of $CdSO_4$. The sludge was incorporated into the soil at a rate of 10 g/kg soil and the test plants
were grown to the commercial harvest stage. This experiment led to
the ranking of the plants according to their tolerance to Cd, as
judged by yield and leaf Cd levels (Table 2). The addition of 10 mg
Cd per kg soil at pH 7.5 was sufficient to reduce yields of spinach
(*Spinacia oleracea* L.), soybean (*Glycine max* L.), and curly cress

TABLE 1

Solution Culture Concentration of Cd Causing a 25% Reduction in Growth,
Corresponding Leaf Cd Concentration, and Visible Symptoms

Crop species	Cd concentration at a 25% growth decrement (mg/liter)	Corresponding leaf Cd concentration[a] (mg/kg)	Visible leaf symptoms
Beet (*Beta vulgaris* L.)	0.05	150	Wilt
Field bean (*Phaseolus vulgaris* L.)	0.08	9	Wilt
Turnip (*Brassica rapa* L.)	0.10	160	Chlorosis
Lettuce (*Lactuca sativa* L.)	0.25	150	Chlorosis
Corn (*Zea mays* L.)	0.35	200	Chlorosis
Pepper (*Capsicum frutescens*)	0.50	60	Chlorosis
Barley (*Hordeum vulgare* L.)	0.75	50	None
Tomato (*Lycopersicon esculentum* L.)	1.00	175	Chlorosis
Cabbage (*Brassica oleracea* L.)	4.00	600	None

[a]Dry weight basis.
Source: Adapted from Ref. 8.

TABLE 2

Soil and Plant Tissue Cd Levels Associated with a 25% Decrement in Yield

Crop species	Soil Cd level causing a 25% decrement in yield[a] (mg/kg)	Leaf Cd content at a 25% decrement in yield[a] (mg/kg)
Spinach (*Spinacia oleracea* L.)	4	75
Soybean (*Glycine max* L.)	5	7
Curly cress (*Lepidium sativum* L.)	8	80
Lettuce (*Lactuca sativa* L. var. *longifolia*)	13	90
Corn (*Zea mays* L.)	18	35
Carrot (*Daucus carota sativa* L.)	20	30
Turnip (*Brassica rapa* L.)	28	120
Field bean (*Phaseolus* spp.)	40	15
Wheat (*Triticum aestivum* L.)	50	35
Radish (*Raphanus sativus* L.)	96	75
Tomato (*Lycopersicon esculentum* L.)	160	125
Squash (*Cucurbita bepo* L.)	160	70
Cabbage (*Brassica oleracea* L.)	170	160
Swiss chard (*Beta vulgaris* var. *cicla*)	250	150
Upland rice (*Oryza sativa* L.)	>640	3
Paddy rice (*Oryza sativa* L.)	17	—

[a]Dry weight basis.
Source: Adapted from Ref. 9.

(*Lepidium sativum* L.) by 25% or more. In contrast to these relatively low threshold concentrations, concentrations of $\geq$160 mg Cd
per kg soil were required for similar yield reductions in tomato
(*Lycopersicon esculentum* L.), squash (*Cucurbita bepo* L.), cabbage
(*Br. oleracea* L.), Swiss chard (*B. vulgaris* L.), and wetland rice
(*Oryza sativa* L.). Rice grown in the greenhouse under upland conditions was much more sensitive to soil Cd, with a concentration of
17 mg Cd per kg soil reducing yields by 25% [10].

The effect of soil type on Cd uptake and threshold leaf and
soil Cd levels is shown for eight soils in Table 3 [11]. Leaf Cd
data are reported for soil addition rates of 0, 5, 10, and 20 mg
Cd per kg for each soil. Also included are the threshold concentrations associated with a 25% decrement in shoot weight. These

TABLE 3

Cadmium Accumulation by Lettuce (*Lactuca sativa* var. *longifolia*)
in Relation to Cd Treatment Rate of Acid and Calcareous Soils
and Threshold Leaf and Soil Cd Concentrations
(25% Yield Decrement)

Soil series	Soil pH	Leaf Cd content[a] (mg/kg) for Cd rate (mg/kg) of				Threshold Cd level (mg/kg)	
		0	5	10	20	Leaf[a]	Soil
Altamont	4.7	0.6	21	42	64	150	60
Hanford	5.0	0.8	19	32	48	200	30
Redding	5.4	0.9	21	40	77	200	100
San Miguel	5.7	0.8	29	50	71	220	160
Arizo	7.4	1.3	42	42	71	90	10
Domino	7.5	0.9	88	88	111	120	50
Simona	7.7	0.5	52	52	71	85	30
Holtville	7.7	3.0	62	62	99	65	10

[a]Dry weight basis.
Source: Adapted from Ref. 11.

data show leaf Cd values for the 10 mg Cd per kg soil treatment, ranging from 32 for the Hanford soil to 88 for the Domino soil. Threshold leaf Cd values ranged from 65 to 220 mg/kg, with the higher values being observed for plants grown in the acid soils. The threshold soil Cd rates (associated with a 25% yield depression) were also quite variable, with values varying from 10 to 160 mg/kg soil. These data clearly demonstrate that diagnostic criteria for Cd toxicity are highly dependent upon the crop species as well as upon soil type.

The experience with Cd availability under field conditions indicates that for a given application rate, Cd uptake by the test plant is not as great as that observed with pot tests [12]. With potted soils, the plant extends its root system throughout the soil volume treated with Cd. In the field, root systems probably extend beyond the depth of soil treated with Cd; consequently, the field-grown plants absorb less Cd. The long-term field experiment conducted by Chang et al. [13] showed that the Cd content of barley (Hordeum vulgare L.) increased according to the annual addition rate of Cd added as sewage sludge, but not according to the cumulative rate. Over the 6-year period of sewage sludge additions, a total of 50 kg Cd per hectare was applied, yet the Cd uptake was minimal (<0.1 mg Cd per kg grain on a dry weight basis) and, as expected, no yield depression occurred. In the midwestern part of the United States Jones et al. [14] followed Cd uptake by corn (Zea mays L.) grown in small plots treated with different amounts of sewage sludge each year for several years. The highest leaf Cd value was 28 mg/kg leaf material (dry weight basis); however, no reduction in yields occurred. They found leaf Cd concentrations to parallel the annual Cd addition rates but not the cumulative rates. Additional details on Cd availability in these plots were published by Hinesly et al. [15]. Additional information on the availability of sludge-borne Cd under field conditions is presented in the review paper by Sommers [16].

Observations of the availability of native Cd in soils to vegetable crops are limited. Lund et al. [17] assessed Cd availability in seven uncultivated soils using several crop species. The 4 N HNO_3 extractable Cd varied from 0.02 to 22 mg/kg soil and leaf Cd values for Swiss chard (*B. vulgaris* var. *cicla*) varied from 0.6 to 82 mg/kg leaf material (dry weight basis). The correlation between leaf Cd and soil Cd (r = 0.92) was highly significant.

Cadmium in soil solution of acid soils exists as Cd^{2+}, $CdCl^{+}$ and $CdSO_4^{0}$, and in alkaline soils as Cd^{2+}, $CdCl^{+}$, $CdSO_4^{0}$, and, to a lesser degree, $CdHCO_3^{+}$. The Cd^{2+} species accounts for 80-90% of the total Cd in solution, except in soils which have been salinized with Cl^{-} and SO_4^{2-} salts. Under saline conditions the $CdCl^{+}$ and $CdSO_4^{0}$ species account for up to 50% of the total Cd [5,6]. The Cd available to Swiss chard (*B. vulgaris* var. *cicla*) correlated better with the Cd^{2+} activity in soil solutions than with total soluble Cd or free-ion and ion-pair concentrations [5,6]. Thus assessment of the availability of Cd, and undoubtedly other metals, could be refined by chemically speciating trace metals in solution. We have found the GEOCHEM computer program [18] to be satisfactory for speciating Cd, with the calculated Cd^{2+} concentrations corresponding to concentrations measured with the Cd-specific ion electrode [19].

2.3.2. Soil pH

In general, Cd uptake by plants is greatly reduced upon liming an acid soil. The resulting increase in soil pH reduces the solubility of Cd in the soil solution, as well as the bioavailability of soil Cd [6]. Liming a Cd-amended Redding sandy loam increased the pH from 4.0 to approximately 7.0 and reduced soluble Cd in the soil from 0.15 to 0.0035 mg/liter. Leaf Cd contents of Swiss chard (*B. vulgaris* var. *cicla*) grown in the unlimed and limed soil treated at equivalent Cd rates were, respectively, 74 and 8.5 µg Cd per g soil. Similarly, rice (*O. sativa* L.) under "flooded" management (paddy) showed reductions in Cd uptake with increasing pH [20]. Figure 1 illustrates the relationship between Cd treatment rate for the Redding sandy soil at various pH levels and the Cd contents of rice grain. Wheat (*Triticum aestivum* L.) also demonstrated less Cd uptake upon liming of the Redding sandy loam [21]. Reduction in Cd availability due to liming

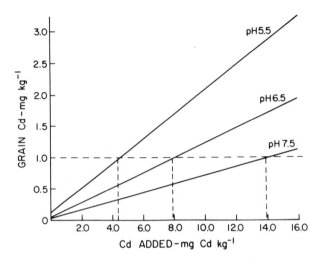

FIG. 1. Predicted content of Cd in rice grain in relation to soil pH and soil Cd of plants grown in the Redding sandy loam treated with Cd ± lime and maintained under flooded conditions. (Adapted from Ref. 20, by permission of the Williams and Wilkins Co., Baltimore, Maryland.)

under field conditions has been discussed by Chaney et al. [22], the Council on Agricultural Science and Technology [23], and Hyde et al. [24] for soybean (*G. max* L.), Swiss chard (*B. vulgaris*), oats (*Avena sativa*), wheat (*T. aestivum*), and corn (*Z. mays*).

The negative effect of increased pH on Cd availability is due in part to reduction of Cd solubility in the soil solution phase. Mahler et al. [25] compared Cd uptake by lettuce, corn, tomato, sweet corn, and Swiss chard grown on four acid and four calcareous soils according to the concentration of Cd in soil solution (Fig. 2). Cadmium uptake as a function of the Cd concentration in soil solution was decidedly higher for plants grown in the acid soils. Root activity influencing Cd absorption may have been stimulated by the lower pH conditions. The greater uptake rate could not be explained upon the basis of the chemical speciation [25]. Additional details of the negative effect of increased pH on Cd availability are contained in recent reviews [1,2,16,26,27].

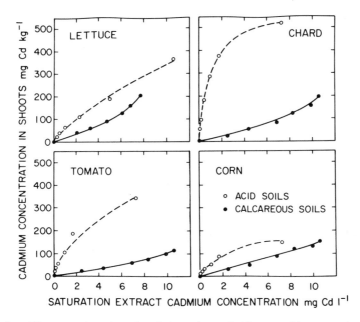

FIG. 2.   Cd concentration in shoots in relation to Cd concentration
of saturation extracts of 4 acid and 4 calcareous soils. (Adapted
from Ref. 25, by permission of the Am. Soc. of Agronomy, Crop. Sci.
Soc. of America, and Soil Sci. Soc. of America).

2.3.3.   Cation Exchange Capacity

The U.S. Environmental Protection Agency recommends disposal rates
for sewage sludge containing more than 2 mg Cd per kg on soils with
pH values above 6.5 according to the cation exchange capacity (CEC)
of the receiving soil [28].   For soils with CEC values of <5, 5-15,
and >15 mEq per 100 g the corresponding limits are 5, 10, and 20 kg
Cd per hectare.   John et al. [29] compared the bioavailability of
Cd in 30 different potted soils treated at a rate of 10 mg Cd per
kg, using radish (Raphanus sativa L.) and lettuce (Lactuca sativa L.)
as indicator plants.   They found the ability of soil to adsorb Cd to
be the most effective parameter reducing Cd availability as judged
by Cd uptake of the test plants.   Haghiri [30] removed organic matter
from a single test soil and then added different amounts of muck to
produce a range of CEC values for the soil-muck mixtures.   The pH

values of the mixtures were held constant at 6.5. The availability of Cd added as $CdCl_2$ was determined by a short-term pot experiment using oats (A. sativa L.) as the test plant. The concentration of Cd in the shoot was negatively correlated with the CEC at a given Cd treatment rate. Miller et al. [31] found that the availability of Cd to soybeans (G. max L.) decreased with increasing CEC in nine soils amended with equivalent rates of $CdCl_2$. In contrast to the above findings, Mahler et al. [11] did not detect a consistent CEC effect on Cd availability using Swiss chard (B. vulgaris) or Romaine lettuce (L. sativa) as test plants for eight acid and calcareous soils treated with sludge pretreated with $CdCl_2$. Hinesly et al. [32] carried out a pot experiment with nine soil mixtures having a range of CEC values (5-15 mEq per 100 g) and treated with sewage sludge at a rate equal to 10 mg Cd per kg or with $CdCl_2$ at the same Cd rate. Corn (Z. mays L.) was used as the indicator plant, with Cd uptake taken as the index of Cd availability. Cadmium uptake was inversely related to the CEC with soil mixtures receiving $CdCl_2$. However, Cd uptake was not influenced by the CEC with soil mixtures treated with sewage sludge containing Cd. The conflicting observations of CEC affecting Cd availability point to the need for further research.

2.3.4. Redox Potential

Increased availability of soil Cd to rice plants growing in paddies was directly related to the number of days that the paddies were drained prior to harvesting the crop [33,34]. This observation suggested that Cd solubility, and hence availability, was strongly dependent upon the redox potential of the soil. Reddy and Patrick [35] conducted a short-term experiment with rice seedlings growing in soil suspensions having controlled levels of redox potential and pH to demonstrate their combined effect on Cd solubility and availability. Decreasing the redox potential from 400 to -200 mV greatly decreased the solubility and availability of Cd in soil suspensions adjusted to different pH levels .Fig. 3).

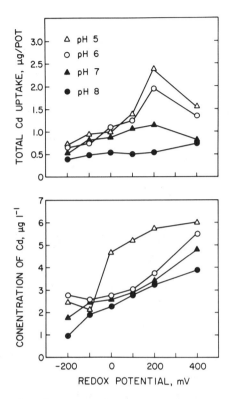

FIG. 3.   Effect of redox potential and pH on Cd uptake by rice
plants and on the Cd concentration of the soil-water suspension.
(Adapted from Ref. 35, by permission of the Am. Soc. of Agronomy,
Crop Sci. Soc. of America, and Soil Sci. Soc. of America).

Bingham et al. [10] also demonstrated the marked effect of
redox potential on Cd solubility and availability to rice plants by
growing rice in flooded (paddy) and nonflooded (upland) soil treated
with low to excessive amounts of Cd.   A 25% decrement in grain production was associated with soil Cd treatments of 17 and 320 mg/kg
soil, respectively, for the nonflooded and flooded management systems (Fig. 4).   The concentrations of Cd in the soil water phase
(saturation extracts of soil with water) of soil treated at the 17 mg
Cd per kg soil rate were, respectively, 0.003 and 2.01 mg Cd per liter
for the flooded and nonflooded systems.   Both leaf and grain contained
lower concentrations when rice was grown under flooded conditions.

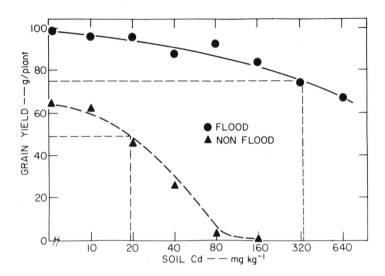

FIG. 4.   Grain yield of rice grown under flooded and nonflooded
conditions in relation to soil Cd concentration.   (Adapted from
Ref. 10, by permission of the Am. Soc. of Agronomy, Crop Sci. Soc.
of America, and Soil Sci. Soc. of America).

Chemical analysis of soil water collected at harvest time from
the flooded and nonflooded soil pots showed SO_4^{2-} to be at a nonde-
tectable range (<1.0 mg/liter) under flooded conditions and ranging
up to 19.2 mg/liter under nonflooded conditions.   The lack of SO_4^{2-}
under low redox conditions suggests that Cd precipitated as CdS and
was thus less available to rice.   However, the increased Mn^{2+} and
Fe^{2+} concentrations associated with flooded soils may have also
curtailed uptake of Cd by the plant [35].

2.4. Cd Toxicity Relative to Cu, Ni, and Zn

Our laboratory group has tested the phytotoxicity of Cd, Cu, Ni, and
Zn alone and in combination with acid and calcareous soils.   For
example, Mitchell et al. [36] investigated the phytotoxicity of
these metals individually by incorporating a sewage sludge (1% addition rate) enriched with low to phytotoxic amounts of $CdSO_4$, $CuSO_4$,

$NiSO_4$, or $ZnSO_4$ in an acid soil as well as a calcareous soil and growing lettuce (*L. sativa* L.) and wheat (*T. aestivum* L.) to maturity. Depending on the soil and plant species, they found Cd to be 2-20 times more toxic than the other metals. In general, the order of toxicity was found to be Cd > Ni > Cu > Zn.

Interactive effects of these metals were evaluated in pot tests with wheat as the test plant for an acid soil with factorial treatments of metals with and without liming [37]. Wheat grain yield (Y in g/plant) for the acid soil was found to be related to metal additions (mg/kg soil) as follows:

$$Y = 3.71 - 0.0036(Zn + 1.4\ Cu + 2.1\ Ni + 4.0\ Cd)$$

This shows Cd to be approximately four times more toxic than Zn. Under limed conditions only Cd and Cu were phytotoxic at the levels tested (up to 80 mg Cd per kg, 200 mg Cu per kg, 80 mg Ni per kg, and 200 mg Zn per kg). Under these conditions Cd was 4.5 times more toxic than Cu.

Relative toxicity coefficients are used to estimate the "zinc equivalent" [37-41] or the potential toxicity of metals to agricultural crops cultivated on soils with pH values of at least 6.5. Accordingly, toxicity is probable when the sum of Zn + 2Cu + 8 Ni (mg/kg soil) exceeds 250 mg/kg for soils with pH values of at least 6.5. Cadmium could be incorporated into the Zn equivalent expression provided that its toxicity coefficient was established. However, metal toxicities vary according to certain soil properties, pH, for example, and plant species. Consequently, the zinc equivalent, or metal toxicity coefficient for that matter, should be used judiciously. The toxicity of these metals added to soil would increase under acid soil conditions.

2.5. Corrective Measures

Cadmium is quite immobile in soil and when materials containing Cd are applied to the soil surface, essentially all of the Cd applied

remains in the depth of incorporation. Means to decontaminate this
layer include removal of the contaminated layer, removal of the Cd
from the layer, or covering the contaminated layer. Cadmium can be
complexed with commercially available chelate—for example, ethylene-
diaminetetraacetic acid (EDTA)—and leached below the root zone.
Although this procedure has been used successfully in Japan, it is
expensive and may lead to contamination of the goundwater. Removal
of the contaminated layer or covering the layer with noncontaminated
soil is also effective, but the costs may be prohibitive.

Techniques to minimize the amount of Cd which may enter the
human food chain from contaminated soils include growth of nonfood
chain crops, growth of crops used exclusively as animal feed and
that absorb relatively small amounts of Cd, and growth of crops
which make up only a small percent of the human diet. In general,
crops grown on acid soils absorb more Cd than crops grown on neutral
or calcareous soil. Therefore, liming acid soils to increase the
soil pH is an effective means to reduce the amount of Cd absorbed.

3. COPPER

3.1. Forms in Soil

Naturally occurring copper minerals in soils include oxysulfates,
carbonates, phosphates, oxides, and hydroxides. Copper sulfides may
form in poorly drained or submerged soils where reducing conditions
exist. Copper minerals are normally too soluble to persist in freely
drained agricultural soils [42]. In metal-contaminated soils, however, the chemical environment may be controlled by nonequilibrium
processes that lead to the accumulation of metastable solid phases.
Recent evidence suggests that covellite (CuS) or chalcopyrite
($CuFeS_2$) may exist as thermodynamically stable minerals in both
oxidized and reduced Cu-contaminated soils [43].

Trace amounts of Cu occur as discrete sulfide inclusions in
silicates and isomorphously substitute for structural cations in

phyllosilicates [44]. Charge-unbalanced clay minerals nonspecif-
ically adsorb Cu while oxides and hydroxides of Fe and Mn demonstrate
high specific Cu affinity [45]. High molecular weight organic components act as solid adsorbents; low molecular weight organic materials tend to form readily soluble Cu complexes [46].

The compositional complexity of soils limits the ability to
quantitatively partition total soil Cu into unambiguous chemical
forms. Sequential extraction procedures indicate that the bulk of
nonstructural Cu resides in organic matter and in Mn and Fe oxides.
Readily soluble and exchangeable Cu, considered the forms that are
available for plant uptake, typically comprise less than 5% of total
soil Cu [47,48]. Application of Cu-containing sewage sludge or
inorganic Cu salts enhances the concentrations of soil Cu that are
extractable by milder reagents [49,50], suggesting that Cu may exist
in more labile chemical forms in Cu-polluted soils.

Uncertainties in thermodynamic equilibrium constants and lack
of experimental techniques have restricted precise determination of
Cu speciation in soil solutions. In the system $Cu-CO_2-H_2O$ the predominant soluble species shifts from free ionic Cu^{2+} to $CuCO_3^0$ to
$Cu(OH)_2^0$ with increasing pH. Free ionic Cu occurs as the hydrated
ion $[Cu(H_2O)_6]^{2+}$ and commonly forms coordination complexes by ligand
substitution of the aquo groups [51]. The preponderance of soluble
Cu in soil solutions occurs in organic complexes [52,53], although
free ionic Cu^{2+} may be significant at low soil pH values [45].
Copper solubility is enhanced under acid conditions and normally
decreases under reducing conditions [54]. The solution species and
oxidation states are largely determined by the stabilizing effects
of complexing ligands [51].

3.2. Availability to Plants

Root absorption of soil solution Cu is the primary means by which
Cu enters plants. Hydroponic experiments demonstrate that plant
tissue concentrations, total uptake, and toxicity symptoms of Cu

tend to increase with increasing Cu concentration in the solution
bathing the plant roots. The chemical form of soluble Cu is of
greater importance to plant uptake processes than is the total
soluble concentration. Addition of the metal chelators EDTA and
diethylenetriaminepentaacetic acid (DTPA) to Cu-containing nutrient
solutions enhanced yields and decreased Cu concentrations and toxicity symptoms in hydroponically grown plants [55,56]. In contrast,
EDTA and DTPA additions enhanced Cu concentrations in barley (H.
vulgare) grown in soil [57].

Net complex charge governed the uptake of several Cu complexes
by excised barley roots (H. vulgare), although maximum absorption
occurred with the Cu^{2+} form [58]. In general, Cu must be released
from organic chelates before it can be absorbed by roots [59].
Relative yield and tissue Cu concentrations of maize (Z. mays) was
significantly correlated with the activity of Cu^{2+} in hydroponic
and field experiments [56,60].

These data suggest that chelating substances both enhance Cu
solubility and inhibit Cu absorption by roots, with the former process dominating in soils. Copper complexation may be further promoted in the immediate root environment by root excretion of soluble
organic chelates. Soluble Cu-organic complexes maintain elevated
levels of potentially available Cu at root surfaces where free ionic
Cu^{2+} is absorbed after decomposition. In lower pH environments Cu^{2+}
is directly available for absorption by roots.

The effective availability of soil Cu to plants is governed by
dynamic interactions between soil factors that influence the supply
of Cu to roots and plant factors that determine Cu absorption,
entrance into the symplast, and redistribution within the plants.
Physical models have not been developed that accurately describe
soil transport processes and root absorption in high Cu soils.
Statistical correlations between soil parameters and plant uptake
or growth responses are the primary means of assessing Cu bioavailability.

Crops on sewage sludge-amended soils often develop elevated Cu
concentrations in certain plant tissues [61], although the response

is not universal [62]. Yields on such soils are normally stimulated by sludge-borne plant nutrients rather than depressed by heavy metal components. Prolonged or very heavy applications of sewage sludge may eventually lead to crop injury, although the phytotoxic effects usually cannot be attributed to Cu alone [63].

Long-term field application of $CuSO_4$ to near-neutral soils did not enhance Cu concentrations or reduce yields of maize (Z. mays) and soybean (G. max) [50,64]. Copper applied at equivalent rates either in Cu-enriched hog manure or as $CuSO_4$ produced similar Cu uptake responses in maize while yields were unaffected [65]. Copper toxicity damage to young grape vines (Plasmopara viticola) resulted from soil contamination by Cu-salt fungicides [66]. Toxicity was promoted by soil acidity and low cation exchange capacity. Extractable-Cu enrichment occurred only in the surface layers of the soils; hence deep-rooted crops were unaffected. Single-time soil applications of $CuSO_4$ and $Cu(OH)_2$ at high rates reduced yields and increased tissue Cu concentrations in snapbean (P. vulgaris) [67] and maize (Z. mays) [60]. The residual toxicity of heavy loading rates of Cu in soils does not decrease rapidly [67,68], although long-term studies have not been made.

Environmental and nutritional factors may influence Cu phytotoxicity. Copper toxicity to paddy rice (O. sativa) was more severe when the plants were irrigated with cold water than with warm water. Microbial activity was retarded in the cold soil and failed to generate reductive soil conditions that would precipitate Cu sulfides from solution [69]. Copper concentrations in grasses may be enhanced by adding N to Cu-contaminated soils [70,71]. High soil levels of Cu reduced the P content of maize (Z. mays) [50,60].

Crop species demonstrate differential partitioning of Cu in plant parts when grown in high Cu soils [61,62]. Under toxic soil conditions Cu tends to be preferentially accumulated in roots [56, 63]. Tissue Cu concentrations were correlated with incipient yield reductions in five crop species [72]; however, plant response to Cu excess is undoubtedly governed by more complex causal factors [56,

73]. The effects of excess Cu on plant metabolic systems are extensively discussed elsewhere [74].

3.3. Symptomology and Diagnosis

Typical symptoms of Cu phytotoxicity include chlorosis, reduced
shoot growth, abnormal root development, and wilting. Chlorosis
appears to result from Cu-induced Fe deficiency, although other
causal factors exist [74]. Excess Cu reduces root length and the
number of root hairs and secondary roots. Root cap development is
abnormal and root elongation may cease [75]. Subsequent inability
to absorb nutrients and water may produce crop stunting and wilting.
The symptoms may be masked or enhanced by other factors, including
both plant and soil characteristics [74]. Copper toxicity symptoms
for specific crops are summarized elsewhere [76-78].

National guidelines for maximum soil additions of Cu in sewage
sludge range from 30 to 280 kg Cu per hectare. The variation reflects differences in crops, soils, climate, and political attitudes
in different countries [63]. Standardized soils tests were developed
to predict Cu deficiency and have been adapted for evaluating toxic
soil conditions [78]. Recent trends are to use saturation extract
or sequential extraction procedures and interpret the results based
on correlative studies of crop response [5,49]. Correlations may be
improved by including additional soil or plant parameters [60].

Copper toxicity symptoms are usually expressed when plant
tissue concentrations exceed 20-30 mg Cu per kg dry matter. Data
for selected crops appear elsewhere [72,76,78].

3.4. Corrective Measures

Excessive foliar applications for Cu-containing pesticides may produce acute Cu toxicity. This risk can be avoided by using proper
application rates or by shifting to organic and organometallic fungicides that do not contain Cu.

Copper phytotoxicity results primarily from excessive soil levels of available Cu and is promoted by soil acidity. Liming enhanced maize yields in a $CuSO_4$-contaminated field soil [60]. Copper-induced Fe deficiency was eliminated by soil application of Fe-EDTA and lime or by foliar application of $FeSO_4$ [74]. Agricultural soils are normally contaminated by surface applications of Cu sources. Since Cu is relatively immobile in soil, almost all of the Cu added to soil will remain in the surface layers [65,79]. Deep plowing of such soils may reduce toxicity where subsoils contain free $CaCO_3$ or high organic matter [66]. Increasing the organic matter content of Cu-affected soils may decrease the incidence of toxicity by adsorption of soluble Cu on organic substrates, by converting Cu^{2+} to less plant-available complexed forms, or by inducing Cu mobility and leaching by forming soluble Cu-organic complexes [60,66]. In a glasshouse experiment, addition of tetraethylenepentamine to high-Cu soils reduced Cu uptake and enhanced germination and the dry matter yield of maize (Z. mays) [80]. These results were attributed to the reduction of soluble Cu concentrations by adsorption of the Cu-tetraethylenepentamine complex to clays and to limited root absorption of the complex.

Managing soil pH and limiting N applications may control Cu concentrations in crops in cases of moderate Cu excess [71]. Reclamation of high-Cu soils by harvesting Cu-accumulating plants appears unfeasible [68]. Physically removing or burying Cu-contaminated topsoil is an effective reclamation strategy [81].

4. ZINC

4.1. Forms in Soil

Zinc materials in soils may include oxysulfates, carbonates, phosphates, silicates, oxides, and hydroxides. These minerals are metastable in well-drained agricultural soils; hence they are unlikely to be present in substantial quantities [42]. Sphalerite (ZnS)

appears to be the thermodynamically favored form in both reduced and
oxidized soils [43]. Minor associations of Zn with P and Cl were
present in a reduced, heavy metal-contaminated sediment, suggesting
that relatively soluble Zn minerals may occur in metal-rich soils
[43].

Zinc isomorphously substitutes for other cations in silicate
minerals and may be occluded or coprecipitated with hydrous oxides
of Mn and Fe. Phyllosilicates, carbonates, hydrous metal oxides,
and organic matter strongly adsorb Zn; both specific and nonspecific
adsorption sites may be present [82].

The compositional complexity of soils limits the ability to
quantitatively partition total soil Zn into distinct solid phases.
Sequential extraction procedures suggest that Zn resides primarily
in nonextractable forms in uncontaminated soils; the extractable
portion is associated mainly with Al and Fe hydrous oxides [48,83].
The relative percentage of nonextractable Zn decreases and the proportion of more readily extractable Zn increases in Zn-contaminated
soils [49,84,85]. The data suggest that Zn may exist in more labile
chemical forms in high-Zn soils.

The chemical species of Zn in soil solutions has not been
unambiguously established. Free ionic Zn occurs as the hexaquo ion
$[Zn(H_2O)_6]^{2+}$ and commonly forms coordination complexes or ion pairs
by ligand substitution of the H_2O groups. In the system $Zn-CO_2-H_2$),
the free ionic form predominates below pH 7.7. The monovalent cation
$ZnOH^+$ is the major species at pH 7.7-9.1 and $Zn(OH)_2^0$ or $Zn(OH)_4^{2-}$
dominates above pH 9.1 [42]. The ion pairs $ZnSO_4^0$, $ZnHCO_3^+$, $ZnHCO_3^0$
or $Zn(HCO_3)_2^0$, and $ZnHPO_4^0$ may be important species, depending on pH
and solution composition [42,86]. Organic complexes accounted for
28-99% of the total soluble Zn in displaced soil solutions of less
than 0.2 mg/liter [87]. Qualitatively similar data have been established for high-Zn soil solutions [45]. Zinc solubility is enhanced
under acid conditions and by Zn complexation with low molecular
weight organic ligands [42]. Reducing conditions may decrease Zn
solubility by formation of insoluble ZnS [43]; however, the enhanced

production of organic chelates and dissolution of Zn-containing metal
oxides appear to predominate in saturated or submerged soils, causing
soluble Zn concentrations to increase [54,88].

4.2. Availability to Plants

Zinc phytotoxicity usually results from exposure of plant roots to
excessive levels of soil solution Zn. Dissolution and desorption of
Zn from solid sources and transport of soluble Zn from the bulk soil
to root surfaces determine Zn concentrations in the rhizosphere.
Zinc transport through the soil occurs by convection and diffusion,
with the latter process dominating in low-Zn soils [89,90]. Convec-
tive transport may be more important in high-Zn soils where soluble
Zn concentrations are relatively stable. If the rate of supply
exceeds the root absorption rate, concentrations of Zn that exceed
bulk soil solution values may develop near the root. Soil mass flow
transport processes may therefore enhance Zn phytotoxicity in cases
of oversupply. The role of transport processes in Zn-contaminated
soils is not well understood.

Zinc mobility in soils is enhanced by the presence of natural
or synthetic chelators. The increase in soluble Zn concentrations
caused by chelation compensates for reductions in mobility resulting
from larger molecular size [90]. Hydroponic experiments demonstrate
that plant tissue Zn concentrations, total uptake, and toxicity
symptoms are positively correlated with the Zn concentration of the
solution bathing the plant roots [91,92]. Zinc absorption by intact
plant roots appears to be metabolically mediated, at least at low
soil solution Zn concentrations [93].

Additions of organic chelates to nutrient solutions decreases
Zn concentrations in tissues of hydroponically grown crops [94,95];
the chemical form of the chelate influences the relative concentration decrease. Similar additions of chelates to soils enhance Zn
concentrations in plants [57]. Correlative studies suggest that
free ionic Zn^{2+} is the chemical species that is preferentially

absorbed by plant roots [94]. These data suggest that chelates
promote Zn solubility in soils and that this process compensates
for reduced plant availability of the Zn in chelated form.

Zinc concentrations in field-grown crops are enhanced by soil
applications of inorganic Zn salts [50] or Zn-containing sewage
sludges [13,96]. Glasshouse experiments showed that Zn concentrations
in four crop species increased linearly with increasing extractable
soil Zn concentrations [97]. In the field, however, tissue concentrations tend to become asymptotic to a crop-specific maximum value at
high extractable soil Zn concentrations or high sludge application
rates [98,99]. Zinc concentrations in maize (Z. mays) decreased in
successive crops after sludge applications to a neutral-reaction soil
were stopped [100], while Zn concentrations remained high in snapbean
(P. vulgaris) grown for several years on a sludge-amended acid soil
[98]. Plant availability of Zn in soil-applied materials of equivalent total Zn contents decreased in the order $ZnSO_4$ > sewage sludge >
garbage compost [101], indicating that the original form of contaminating metal influences toxicity potential.

Most field studies of soil contamination by Zn-enriched sewage
sludge fail to demonstrate crop yield reductions or phytotoxicity
symptoms; however, long-term loading at high rates may eventually
produce plant damage [63]. Inorganic Zn salts appear to be more
available to plants. Single applications of Zn as $ZnSO_4$ caused yield
reductions in crops grown on acid soils [67,102], whereas multiyear
additions of $ZnSO_4$ to near-neutral soils had no effect [50,64].

Environmental and nutritional factors may influence Zn phytotoxicity. Liming of sludge-amended soils reduced Zn concentrations
in rye (Secale vulgaris) and maize (Z. mays) [13,103]. Warm soil
temperatures enhanced Zn concentrations in several vegetable crops
grown in sludge-amended soils [104]. Plant nutrients in soil-applied
sewage sludge normally stimulate crop yields, which can ameliorate Zn
phytotoxicity by distributing Zn throughout a larger plant volume
[105]. Ion interactions are difficult to unambiguously demonstrate
in the field. Plant uptake of Zn is normally inhibited by high

concentrations of other cations in nutrient solutions or potted soils
[106]. In order to produce this result, the competing ions must be
present in high concentrations. Since these ions are commonly toxicants, their substitution for Zn does little to alleviate plant
injury [37].

Toxic Zn levels in agricultural soils normally result from
surface applications of high-Zn materials. The redistribution of
surface-applied Zn is usually restricted to the upper 15-30 cm of
the soil [50,79]. Deep-rooted crops may circumvent the deleterious
effects of excess soil Zn by establishing roots in uncontaminated
subsoil, as occurs in Cu-contaminated soils [66].

Crop species differ in the partitioning of Zn among plant
organs when grown in identical soils [61]. Twenty cultivars of soybean (G. max) demonstrated a range of Zn uptake and yield responses
to high soil Zn levels. Zinc was preferentially accumulated in the
roots of four soybean cultivars relative to the shoots [107]. The
cultivars demonstrated differential absorption and translocation of
Zn. The relationship between tissue Zn concentrations and soil Zn
depended on soil pH and cultivar. Scion genotype controlled the
relative Zn tolerances of two soybean varieties, while rootstock
governed Zn absorption and translocation [108]. Differences in
foliar Zn concentrations were not responsible for differential Zn
tolerance. The effect of excess Zn on plant metabolism is discussed
in several recent reviews [77,109].

4.3. Symptomology and Diagnosis

Typical symptoms of Zn phytotoxicity include chlorosis and growth
retardation, resulting in a stunted appearance. The chlorotic
response has been attributed to interference with Fe metabolism,
although leaf Fe concentrations may not be depressed [92]. Chlorosis
may be independent of yield reduction [109]. Zinc toxicity symptoms
in specific crops are summarized elsewhere [77,93,107,109,110].

National guidelines for maximum soil additions of Zn in sewage
sludge vary from 120 to 560 kg Zn per hectare. The values reflect
differences in crops, soils, climates, and political attitudes [63].
Single-soil extraction tests have been proposed to evaluate Zn toxicity potential; correlation equations that include additional soil
parameters usually provide better predictions of plant responses
[111]. Recent trends are to use saturation extract or sequential
extraction procedures [5,99].

Phytotoxicity symptoms or yield reductions caused by Zn excess
are often correlated with plant tissue Zn concentrations [72,109].
Foliar analysis may be impractical as an absolute indicator of Zn
phytotoxicity, since the mathematical relationship between tissue
concentrations and response varies with genotype, plant part, stage
of growth, and growing conditions [107]. In many cases, however,
foliar tissue concentrations in excess of 200 mg Zn per kg dry matter
signal potential phytotoxicity [72,109,110].

4.4. Corrective Measures

Liming of Zn-contaminated soils reduced Zn concentrations in field-
grown crops [96,103]. Application of NaOH or $Ca(OH)_2$ reduced Zn
toxicity in vegetable crops grown on high-Zn peat soils [110]. Zinc-
induced Fe deficiency can be eliminated by soil or foliar application
of Fe-chelates or $FeSO_4$ [110]. Reclamation of high-Zn soils by harvesting Zn in crops appears unfeasible owing to limited uptake of Zn
[112]. Physically removing or burying Zn-contaminated topsoil is an
effective reclamation strategy [81].

5. MANGANESE

5.1. Forms in Soil

Soil manganese is found in three oxidation states: +2, +3, and +4.
Most of the soil Mn is associated with primary minerals or with

secondary Mn oxides. In soils total Mn averages about 500-900 mg/kg
[113].

The solubility of the Mn^{4+} form is extremely low, while trivalent Mn is highly unstable in the soil environment. Most of the
Mn in solution is expected to be present as Mn^{2+}, while in well-
aerated soils most of the Mn in the solid phase is present in the
IV oxidation state, as oxides. In poorly aerated soils this Mn is
slowly reduced by microbial metabolism and then enters the soil
solution, where it is highly mobile.

Manganese will associate with insoluble organic materials in
soil, although this appears to be a less important sink for Mn than
for other transition metals such as Cu. Godo and Reisenauer [114]
found that although Mn^{2+} solubility increased significantly at low
pH, the uptake of Mn^{2+} by the plant decreased. Increased solubility
appeared to be due, at least in part, to reaction with root organic
exudates, suggesting that soluble Mn-organic complexes may be important in field soils. Most of the Mn in soil solution is present as
free Mn^{2+}, and some of it is associated with organic ligands [115].

5.2. Availability to Plants

The uptake of Mn appears to compete with that of some of the other
cations, especially Ca and Mg, which are also divalent [116].

Both water-soluble and exchangeable Mn concentrations are used
to predict which soils contain concentrations of Mn potentially toxic
to plants [117]. Because of the importance of reduction in solubil-
izing Mn, reductants have been used for extraction of Mn as an index
of plant availability [118]. The extraction of Mn^{2+} in DTPA has been
used to estimate Mn availability.

Manganese reportedly associates with soluble organic molecules
in the xylem [119], although others have reported no such association
[120]. In either case, because of its high mobility in the xylem but
relatively low phloem mobility, Mn tends to accumulate in the older
tissues of the plant. The highest Mn concentrations in the plant

thus develop in the older leaves and the stem, especially in the margins of the tissues [121].

5.3. Symptomology and Diagnosis

Because Mn accumulates in margins of older leaves, it can reach extremely high concentrations (>4000 mg/kg) and eventually cause toxicities which appear first in those tissues. The mechanism of the toxicity effect has been attributed to interference with Fe and Ca metabolism. Excessive solution Mn (>0.5 mg/liter) may decrease both absorption and translocation of zinc [122]. Also, the accelerated activity of indole acetic acid oxidase may result in a deficiency of indole acetic acid [77,123,124]. The plant growth pattern may reflect this hormonal deficiency.

Manganese may precipitate as a mixed oxide in plant leaf tissues [125]. Some plants have a remarkable ability to compartmentalize Mn and thus keep it from interfering with important cell processes [126,127]. In this manner the physiological effect of the metal is mitigated, even though total tissue concentrations may be extremely high. The plant may also shed its older leaves and thereby rid itself of excess Mn.

Tissue analysis can be used to diagnose Mn toxicities in plants. Plant tissue concentrations associated with the expression of toxicity range from 173 mg/kg in soybean [128] to over 1000 mg/kg in rice [69]. Since paddy rice is produced in flooded soils, the need for Mn tolerance is evident. Soil analysis is less reliable because of the large effect environment has on the solubility and hence availability of Mn. However, the amount of Mn extracted from soils by DTPA is often highly correlated with plant uptake.

5.4. Corrective Measures

Manganese toxicity frequently occurs where total soil Mn is moderate to high, soil pH is low, and soil oxygen levels are low (i.e., reducing conditions). To alleviate the adverse effects of these conditions, soil pH can be increased by liming and efforts can be made to improve soil drainage, decrease water inputs, or improve soil structure.

There are substantial differences in the tolerance of different varieties of a given species to Mn toxicity (e.g., see Refs. 129 and 130). Selecting for tolerant types may decrease the detrimental impact of excess Mn on plant growth in the field.

Foy [131] proposed that Mn tolerance in plants is associated with (1) the oxidizing powers of plant roots, (2) Mn absorption and translocation rates, (3) Mn entrapment in nonmetabolic centers, (4) high internal tolerance to excess Mn, and (5) the uptake and distribution of Si and Fe.

Much of the adverse effect of Mn is actually due to induced Fe [132,133] or Ca [125] deficiencies. Addition of Fe may decrease Mn uptake and partially alleviate symptoms. Silicon tends to decrease the expression of Mn toxicity (e.g., see Ref. 132).

6. NICKEL

6.1. Forms in Soil

Total nickel concentration in soils ranges from 2 to 750 mg/kg [113, 134]. Nickel toxicity is often associated with serpentine-derived soils [135-137]. Soil acidity increased Ni accumulation by clover [138], and in general Ni availability decreases with increasing soil pH.

Nickel in soils is strongly adsorbed by manganese oxides over a wide pH range and by iron oxides above pH 5.5 [139]. In solution, if sufficient sulfate is present, a substantial fraction of the Ni may be complexed in $NiSO_4^0$ [139].

6.2. Availability to Plants

Plant Ni concentrations are normally within the range of 0.05-5.0 mg/kg [140]. Dixon et al. [141] demonstrated that Ni is an essential component of jack bean urease. More recently, Eskew et al. [142] demonstrated that Ni is essential for soybeans to derive their N from ureides (as products of nitrogen fixation) or simple urea.

Nickel tends to accumulate in roots [143], although transport to tops may be substantial in some cases [144]. Difficulty in completely removing all soil from roots may result in excessive contamination on root surfaces.

Metal-tolerant species have developed means for dealing with excessive levels of available nickel [145,146]. Four types of plant tissue responses to increasing soil Ni have been proposed [146] for plants ranging in type and degree of tolerance to high internal concentrations of Ni.

A high degree of tolerance to high heavy metal concentration has been attributed to strong binding in the cell wall [145]. By their sequestration in the cell wall, metal ions are excluded from potentially sensitive sites within the cell. Plants adapted to high-Ni sites therefore may develop cell walls which are much more effective in complexing Ni.

Plants from areas near a nickel-copper smelter were more tolerant to Ni than were members of the same species grown at some distance from the smelter [147], demonstrating that evolutionary adjustments to excessively high Ni concentrations may occur over reasonably short periods of time. In a similar manner, Johnston and Proctor [136] found that *Festuca rubra* clones from high-Ni serpentine soils in Scotland were more tolerant of Ni than clones from nonserpentine areas. Serpentine soils contained 0.5-0.9 mg Ni per liter of soil water extract (field capacity).

Nickel-induced Fe deficiency has been proposed as a toxicity mechanism of major significance [148]. Nickel is thought to induce typical interveinal chlorosis by inhibiting translocation of Fe from roots to tops [148]. Upper critical levels for Ni concentrations in soil solutions are usually reported in the range of 0.3-0.5 mg Ni per liter [136,149]. In leaf tissue total concentrations at which

plants show demonstrable Ni toxicities range from 25 to 300 mg/kg
[136,140,150]. Specific values depend upon plant species, tissue
age, and type of tissue sampled.

6.3. Symptomology and Diagnosis

Symptoms of Ni toxicity often resemble those of Fe or Zn deficiency
[140]. White or light green striping of leaves may occur in cereals.
With dicotyledonous plants there is often leaf mottling.

6.4. Corrective Measures

Many naturally occurring Ni toxicities are related to serpentine
soils. In these cases liming provides a twofold benefit: (1) Soil
pH is increased, so that Ni solubility is decreased to some extent,
and (2) the Ca added in the lime raises the Ca/Mg ratio and diminishes
the adverse effects of excess Ni [140]. Introduction of plant types
adapted to high-Ni soils could also be practiced to increase establishment, although relatively little has been done to explicitly
select for this plant character.

REFERENCES

1. A. L. Page and F. T. Bingham, *Residue Rev.*, *48*, 1 (1973).

2. A. L. Page, F. T. Bingham, and A. C. Chang, in *Effect of Heavy Metal Pollution on Plants*, Vol. 1 (N. W. Lepp, ed.), Applied Science Publishers, London, 1981, pp. 77-109.

3. G. S. Holmgren, M. W. Meyer, R. B. Daniels, R. L. Chaney, and J. Kubota, *J. Environ. Qual.* (in press).

4. G. Sposito, in *Applied Environmental Geochemistry* (I. Thornton, ed.), Academic Press, New York, 1983, pp. 123-170.

5. F. T. Bingham, J. E. Strong, and G. Sposito, *Soil Sci.*, *135*, 160 (1983).

6. F. T. Bingham, G. Sposito, and . E. Strong, *J. Environ. Qual.*, *13*, 71 (1984).

7. R. J. Mahler, F. T. Bingham, A. L. Page, and J. A. Ryan, *J. Environ. Qual.*, *118*, 694 (1982).

8. A. L. Page, F. T. Bingham, and C. Nelson, *J. Environ. Qual.*, *1*, 288 (1972).

9. F. T. Bingham, A. L. Page, R. J. Mahler, and T. J. Ganje, *J. Environ. Qual.*, *2*, 207 (1975).

10. F. T. Bingham, A. L. Page, R. J. Mahler, and T. J. Ganje, *Soil Sci. Soc. Am. J.*, *40*, 715 (1976).

11. R. J. Mahler, F. T. Bingham, and A. L. Page, *J. Environ. Qual.*, *7*, 274 (1978).

12. A. L. Page and A. C. Chang, in *Proceedings of the Fifth National Conference on Acceptable Sludge Disposal Techniques*, Information Transfer, Rockville, Maryland, 1978, pp. 91-96.

13. A. C. Chang, A. L. Page, J. E. Warneke, M. R. Resketo, and T. E. Jones, *J. Environ. Qual.*, *12*, 391 (1983).

14. R. L. Jones, T. D. Hinesly, E. L. Ziegler, and J. J. Tyler, *J. Environ. Qual.*, *4*, 509 (1975).

15. T. D. Hinesly, R. L. Jones, E. L. Ziegler, and J. J. Tyler, *Environ. Sci. Technol.*, *11*, 182 (1977).

16. L. E. Sommers, in *Sludge—Health Risk of Land Application* (G. Bitton, B. Damron, G. Edds, and J. Davidson, eds.), Ann Arbor Science Publishers, Ann Arbor, Michigan, 1982, pp. 105-140.

17. L. J. Lund, E. E. Betty, A. L. Page, and R. A. Elliot, *J. Environ. Qual.*, *10*, 551 (1981).

18. G. Sposito and S. V. Mattigod, *GEOCHEM: A Computer Program for the Calculation of Chemical Equilibria in Soil Solutions and Other Natural Water Systems*, Kearney Foundation of Soil Science, University of California, Riverside, 1980.

19. G. Sposito, F. T. Bingham, S. S. Yadav, and C. A. Inouye, *Soil Sci. Soc. Am. J.*, *46*, 51 (1982).

20. F. T. Bingham, A. L. Page, and J. E. Strong, *Soil Sci.*, *130*, 32 (1980).

21. F. T. Bingham, A. L. Page, G. A. Mitchell, and J. E. Strong, *J. Environ. Qual.*, *8*, 202 (1979).

22. R. L. Chaney, M. C. White, and P. W. Simon, in *Proceedings of the Second National Conference on Municipal Sludge Management*, Information Transfer Inc., Rockville, Maryland, 1975, pp. 169-178.

23. Council on Agricultural Science and Technology, CAST Report No. 80, Ames, Iowa, 1980.

24. H. C. Hyde, A. L. Page, F. T. Bingham, and R. J. Mahler, *J. Water Pollut. Control Fed.*, *51*, 2475 (1979).

25. R. J. Mahler, F. T. Bingham, A. L. Page, and J. A. Ryan, *J. Environ. Qual., 11,* 694 (1982).

26. A. L. Page, F. T. Bingham, and A. C. Chang, in *Effect of Heavy Metal Pollution on Plants,* Vol. 1 (N. W. Lepp, ed.), Applied Science Publishers, London, 1981, pp. 77-109.

27. J. D. Jastrow and D. E. Koeppe, in *Cadmium in the Environment* (J. O. Nriagu, ed.), John Wiley, New York, 1980, pp. 607-638.

28. U.S. Environmental Protection Agency, *Fed. Reg., 44,* 53438 (1979).

29. M. John, C. J. van Laerhoven, and H. Chuah, *Environ. Sci. Technol., 6,* 1005 (1972).

30. F. Haghiri, *J. Environ. Qual., 3,* 180 (1974).

31. J. E. Miller, J. J. Hassett, and D. E. Koeppe, *J. Environ. Qual., 5,* 157 (1976).

32. T. D. Hinesly, K. E. Redborg, E. L. Ziegler, and J. D. Alexander, *Soil Sci. Soc. Am. J., 46,* 490 (1982).

33. Y. Takijima, F. Katsumi, and S. Koizumi, *Soil Sci. Plant Nutr., 19,* 183 (1973).

34. Y. Takijima, F. Katsumi, and K. Takezawa, *Soil Sci. Plant Nutr., 19,* 173 (1973).

35. C. N. Reddy and W. H. Patrick, Jr., *J. Environ. Qual., 6,* 259 (1977).

36. G. A. Mitchell, F. T. Bingham, and A. L. Page, *J. Environ. Qual., 7,* 165 (1978).

37. F. T. Bingham, A. L. Page, G. A. Mitchell, and J. E. Strong, *J. Environ. Qual., 8,* 202 (1979).

38. J. B. E. Patterson, *Tech. Bull. Minist. Agric. Fish. Food Agric. Dev. Adv. Serv.,* England, *21,* 193 (1971).

39. C. G. Chumbley, *Agric. Dev. Adv. Serv. Minist. Agric.,* England, *10,* 12 (1971).

40. J. Webber, *Water Pollut. Control Fed., 71,* 404 (1972).

41. D. R. Keeney, K. W. Lee, and L. M. Walsh, *Tech. Bull.* No. 88, University of Wisconsin, Madison (1975).

42. W. L. Lindsay, *Chemical Equilibria in Soils,* John Wiley, New York, 1979.

43. F. Y. Lee and J. A. Kittrick, *Soil Sci. Soc. Am. J., 48,* 548 (1984).

44. D. P. Cox, in *Copper in the Environment* (J. O. Nriagu, ed.), John Wiley, New York, 1979, pp. 19-42.

45. M. B. McBride and J. J. Blasiak, *Soil Sci. Soc. Am. J., 43,* 866 (1979).

46. W. F. Pickering, in *Copper in the Environment* (J. O. Nriagu, ed.), John Wiley, New York, 1979, pp. 217-233

47. R. G. McLaren and D. V. Crawford, *J. Soil Sci.*, *24*, 172 (1973).

48. I. H. Elsokkary and J. Lag, *Acta Agric. Scand.*, *23*, 26 (1978).

49. G. Sposito, L. J. Lund, and A. C. Chang, *Soil Sci. Soc. Am. J.*, *46*, 264 (1982).

50. G. L. Mullins, D. C. Martens, S. W. Gettier, and W. P. Miller, *J. Environ. Qual.*, *11*, 573 (1982).

51. J. O. Leckie and J. A. Davis, in *Copper in the Environment* (J. O. Nriagu, ed.), John Wiley, New York, 1979, pp. 89-121.

52. J. F. Hodgson, H. R. Geering, and W. A. Norvell, *Soil Sci. Soc. Am. Proc.*, *29*, 665 (1965).

53. J. F. Hodgson, W. L. Lindsay, and J. F. Trierweiller, *Soil Sci. Soc. Am. Proc.*, *30*, 723 (1966).

54. C. N. Reddy and W. H. Patrick, Jr., *Soil Sci. Soc. Am. J.*, *41*, 729 (1977).

55. S. K. Majunder and S. Dunn, *Plant Soil*, *10*, 296 (1959).

56. J. Dragun, D. E. Baker, and M. L. Risius, *Agron. J.*, *68*, 466 (1976).

57. C. Gonzalez, O. R. F. Goecke, and E. Schalscha B., *Agrochimica*, *16*, 387 (1972).

58. A. J. Coombes, D. A. Phipps, and N. W. Lepp, *Z. Pflanzenphysiol.*, *82*, 435 (1977).

59. B. A. Goodman and D. J. Linehan, in *The Soil-Root Interface* (J. L. Harley and R. S. Russell, eds.), Academic Press, London, 1979, pp. 67-82.

60. Th. M. Lexmond, *Neth. J. Agric. Sci.*, *28*, 164 (1980).

61. R. H. Dowdy and W. E. Larson, *J. Environ. Qual.*, *4*, 278 (1975).

62. R. D. Davis, in *Copper in Animal Wastes and Sewage Sludge* (P. L'Hermite and J. Dehandtschutter, eds.), D. Reidel, Dordrecht, Holland, 1981, pp. 223-241.

63. J. Webber, in *Effect of Heavy Metal Pollution on Plants,* Vol. 2 (N. W. Lepp, ed.), Applied Science Publishers, London, 1981, pp. 159-184.

64. D. C. Martens, M. T. Carter, and G. D. Jones, *Agron. J.*, *66*, 82 (1974).

65. G. L. Mullins, D. C. Martens, W. P. Miller, E. T. Kornegay, and D. L. Hallock, *J. Environ. Qual.*, *11*, 316 (1982).

66. J. Delas, in *Copper in Animal Wastes and Sewage Sludge* (P. L'Hermite and J. Dehandtschutter, eds.), D. Reidel, Dordrecht, Holland, 1981, pp. 136-143.

67. L. M. Walsh, W. H. Erhardt, and H. D. Siebel, *J. Environ. Qual.*, *1*, 197 (1972).

68. A. C. Chang, A. L. Page, J. E. Warneke, and E. Grgurevic, *J. Environ. Qual.*, *13*, 33 (1984).

69. M. Chino, in *Heavy Metal Pollution in Soils in Japan* (K. Kitagishi and I. Yamane, eds.), Japan Scientific Societies Press, Tokyo, 1981, pp. 65-80.

70. J. S. Gladstone, J. F. Loneragan, and W. J. Simons, *Aust. J. Agric. Res.*, *26*, 113 (1975).

71. Y. K. Soon, T. E. Bates, and J. R. Moyer, *J. Environ. Qual.*, *9*, 497 (1980).

72. R. D. Davis and P. H. T. Beckett, *New Phytol.*, *80*, 23 (1978).

73. M. C. White, A. M. Decker, and R. L. Chaney, *Agron. J.*, *71*, 126 (1979).

74. N. W. Lepp, in *Effect of Heavy Metal Pollution on Plants*, Vol. 1 (N. W. Lepp, ed.), Applied Science Publishers, London, 1981, pp. 117-143.

75. W. Savage, W. L. Berry, and C. A. Reed, *J. Plant Nutr.*, *3*, 129 (1981).

76. W. Reuther and C. K. Labanauskas, in *Diagnostic Criteria for Plants and Soils* (H. D. Chapman, ed.), Quality Printing, Abilene, Texas (1966), pp. 157-179.

77. C. D. Foy, R. L. Chaney, and M. C. White, *Annu. Rev. Plant Physiol.*, *29*, 511 (1978).

78. U. C. Gupta, in *Copper in the Environment* (J. O. Nriagu, ed.), John Wiley, New York, 1979, pp. 255-288.

79. D. E. Williams, J. Vlamis, A. H. Pukite, and J. E. Corey, *Soil Sci.*, *137*, 351 (1984).

80. F. Smeulders, J. Sinnaeve, and A. Cremers, in *Plant Nutrition 1978*, Vol. 2 (A. R. Ferguson, R. L. Bieleski, and I. B. Ferguson, eds.), New Zealand DSIR Information Services, Wellington, 1978, pp. 475-490.

81. G. W. Leeper, *Managing the Heavy Metals on the Land*, Marcel Dekker, New York, 1978.

82. L. M. Shuman, in *Zinc in the Environment* (J. O. Nriagu, ed.), John Wiley, New York, 1979, pp. 39-69.

83. L. M. Shuman, *Soil Sci.*, *127*, 10 (1979).

84. S. S. Iyengar, D. C. Martens, and W. P. Miller, *Soil Sci. Soc. Am. J.*, *45*, 735 (1981).

85. M. G. Hickey and J. A. Kittrick, *J. Environ. Qual.*, *13*, 372 (1984).

86. G. Sposito, in *Applied Environmental Geochemistry* (I. Thornton, ed.), Academic Press, New York, 1983, pp. 123-170.

87. J. F. Loneragan, in *Trace Elements in Soil-Plant-Animal Systems* (D. J. D. Nicholas and A. R. Egan, eds.), Academic Press, New York, 1975, pp. 109-134.

88. R. S. Beckwith, K. G. Tiller, and E. Suwadju, in *Trace Elements in Soil-Plant-Animal Systems* (D. J. D. Nicholas and A. R. Egan, eds.), Academic Press, New York, 1975, pp. 135-149.

89. H. F. Wilkinson, J. F. Loneragan, and J. P. Quirk, *Soil Sci. Soc. Am. Proc.*, *32*, 831 (1968).

90. S. M. Elgawhary, W. L. Lindsay, and W. D. Kemper, *Soil Sci. Soc. Am. Proc.*, *34*, 66 (1970).

91. W. E. Rauser, *Can. J. Bot.*, *51*, 301 (1973).

92. J. A. Rosen, C. S. Pike, and M. L. Golden, *Plant Physiol.*, *59*, 1085 (1977).

93. J. C. Collins, in *Effect of Heavy Metal Pollution on Plants*, Vol. 1 (N. W. Lepp, ed.), Applied Science Publishers, London, 1981, pp. 145-169.

94. A. D. Halvorson and W. L. Lindsay, *Soil Sci. Soc. Am. Proc.*, *36*, 755 (1977).

95. P. C. DeKock and R. L. Mitchell, *Soil Sci.*, *84*, 55 (1957).

96. I. L. Pepper, D. F. Bezdicek, A. S. Baker, and J. M. Sims, *J. Environ. Qual.*, *12*, 270 (1983).

97. D. Purves, *Trace-Element Contamination of the Environment*, Elsevier, Amsterdam, 1977.

98. R. H. Dowdy and W. E. Larson, J. M. Titrud, and J. J. Latterell, *J. Environ. Qual.*, *7*, 252 (1978).

99. J. P. LeClaire, A. C. Chang, C. S. Levesque, and G. Sposito, *Soil Sci. Soc. Am. J.*, *48*, 509 (1984).

100. T. D. Hinesly, E. L. Ziegler, and G. L. Barrett, *J. Environ. Qual.*, *8*, 35 (1979).

101. P. M. Giordano, J. J. Mortvedt, and D. A. Mays, *J. Environ. Qual.*, *3*, 394 (1975).

102. C. R. Lee and G. R. Craddock, *Agron. J.*, *61*, 565 (1969).

103. J. V. Lagerwerff, G. T. Biersdorf, R. P. Milberg, and D. L. Brower, *J. Environ. Qual.*, *6*, 427 (1979).

104. P. M. Giordano, D. A. Mays, and A. D. Behel, *J. Environ. Qual.*, *8*, 233 (1979).

105. A. Andersson and K. D. Nilsson, *Swed. J. Agric. Res.*, *6*, 151 (1976).

106. M. K. Hughes, N. W. Lepp, and D. A. Phipps, *Adv. Ecol. Res.*, *11*, 217 (1980).

107. M. C. White, A. M. Decker, and R. L. Chaney, *Agron. J., 71,* 121 (1979).

108. M. C. White, R. L. Chaney, and A. M. Decker, *Crop Sci., 19,* 126 (1979).

109. L. C. Boawn and P. E. Rasmussen, *Agron. J., 63,* 874 (1971).

110. H. D. Chapman, in *Diagnostic Criteria for Plants and Soils* (H. H. Chapman, ed.), Quality Printing, Abilene, Texas (1966), pp. 484-499.

111. A. U. Haq, T. E. Bates, and Y. K. Soon, *Soil Sci. Soc. Am. J., 44,* 772 (1980).

112. A. C. Chang, J. E. Warneke, A. L. Page, and L. J. Lund, *J. Environ. Qual., 13,* 87 (1984).

113. K. B. Krauskopf, in *Micronutrients in Agriculture* (J. J. Mortvedt, P. M. Giordano, and W. L. Lindsay, eds.), Soil Science Society of America, Madison, Wisconsin, 1972, pp. 7-40.

114. G. H. Godo, and H. M. Reisenauer, *Soil Sci. Soc. Amer. J., 44,* 993 (1980).

115. H. R. Geering, J. F. Hodgson, and C. Sdano, *Soil Sci. Soc. Am. Proc., 33,* 81 (1969).

116. E. V. Maas, D. P. Moore, and B. J. Mason, *Plant Physiol., 44,* 796 (1969).

117. F. Adams and J. I. Wear, *Soil Sci. Soc. Am. Proc., 21,* 305 (1957).

118. G. D. Sherman, J. S. McHargue, and W. A. Hodgkiss, *Soil Sci., 54,* 253 (1942).

119. W. Hofner, *Physiol. Plant., 23,* 673 (1970).

120. L. O. Tiffin, *Plant Physiol., 42,* 1427 (1967).

121. J. Vlamis and D. E. Williams, *Plant Sci., 39,* 245 (1973).

122. K. R. Reddy, M. C. Saxena, and U. R. Pal, *Plant Soil, 49,* 409 (1978).

123. C. D. Foy, in *Manganese* (J. Lieben, ed.), National Academy of Sciences, Washington, D.C., 1973, pp. 51-76.

124. C. D. Foy, *Iowa State J. Res., 57,* 339 (1983).

125. W. J. Horst and H. Marschner, *Z. Pflanzenphysiol., 87,* 137 (1978).

126. M. Masui, A. Nukaya, and A. Ishida, *J. Jpn. Soc. Hortic. Sci., 49,* 79 (1980).

127. A. R. Memon, M. Chino, Y. Takeoka, K. Hora, and M. Yatazawa, *J. Plant Nutr., 2,* 457 (1980).

128. K. Ohki, D. O. Wilson, and O. E. Anderson, *Agron. J., 72,* 713 (1980).

129. L. Dessureaux, *Euphytica*, *8*, 260 (1959).

130. J. C. Brown, and F. E. Devine, *Agron. J.*, *72*, 898 (1980).

131. C. D. Foy, *Iowa State J. Res.*, *57*, 355 (1983).

132. W. J. Horst and H. Marschner, *Z. Pflanzenernaehr. Bodenkd.*, *141*, 487 (1978).

133. M. Masui, A. Nukaya, and A. Ishida, *J. Jpn. Soc. Hortic. Sci.*, *45*, 267 (1976).

134. H. J. M. Bowen, *Environmental Chemistry of the Elements*, Academic Press, New York, 1979.

135. W. M. Crooke, *Soil Sci.*, *81*, 269 (1956).

136. W. R. Johnston and J. Proctor, *J. Ecol.*, *69*, 855 (1981).

137. S. Karataglis, D. Babalonas, and B. Kabasakalis, *Phyton (Buenos Aires)*, *22*, 317 (1982).

138. A. C. Oertel, J. A. Prescott, and C. G. Stephen, *Aust. J. Soil Sci.*, *41*, 311 (1954).

139. R. O. Richter and T. L. Theis, in *Nickel in the Environment* (J. O. Nriagu, ed.), John Wiley, New York, 1980, pp. 189-202.

140. A. P. Vanselow, in *Diagnostic Criteria for Plants and Soils* (H. D. Chapman, ed.), Quality Printing, Abilene, Texas, 1966, pp. 302-309.

141. N. E. Dixon, C. Gazzola, R. L. Blakely, and B. Zerner, *J. Am. Chem. Soc.*, *19*, 4431 (1975).

142. D. L. Eskew, R. M. Welch, and E. A. Cary, *Science*, *222*, 621 (1983).

143. L. O. Tiffin, *Plant Physiol.*, *48*, 273 (1977).

144. R. L. Chaney, *Effect of Nickel on Iron Metabolism by Soybeans*, Ph.D. dissertation, Purdue University, Lafayette, Indiana, 1970.

145. J. Antonovics, A. D. Bradshaw, and R. G. Turner, *Adv. Ecol. Res.*, *7*, 1 (1971).

146. A. N. Rencz and W. W. Shilts, in *Nickel in the Environment* (J. O. Nriagu, ed.), John Wiley, New York, 1980, pp. 151-188.

147. R. M. Cox and T. C. Hutchinson, *New Phytol.*, *77*, 547 (1980).

148. W. M. Crooke, J. G. Hunter, and O. Vergnano, *Annu. Appl. Biol.*, *41*, 311 (1954).

149. J. Proctor, *J. Ecol.*, *59*, 827 (1971).

150. R. D. Davis, P. H. T. Beckett, and E. Wollan, *Plant Soil*, *49*, 395 (1978).

6

Metal Ion Toxicity in Man and Animals

Paul B. Hammond and Ernest C. Foulkes
Institute of Environmental Health
Kettering Laboratory
University of Cincinnati College of Medicine
Cincinnati, Ohio 45267-0056

1. INTRODUCTION

In this chapter we describe the salient aspects of metal toxicology.
Obviously, within the available space, it would be impossible to
consider all of the important questions which have been raised in
this field. Nevertheless, some overall picture of how metals behave
in biological systems should emerge. The fact that metals readily
react with biological molecules and especially proteins should not
come as a surprise, given the high reactivity of the metals and the
presence of polyfunctional groups on proteins.

Metals exhibit a broad range of toxicological effects. Some,
such as inorganic lead, adversely affect a great variety of target
organs within the same general range of dosage. In such cases the
most sensitive organ (generally referred to as the "critical organ")
is difficult to perceive. Others, such as vanadium and cadmium,
have a much more limited range of toxic effects, and differences in
the sensitivity of organs are much greater. Thus, in the case of
cadmium, the kidney is clearly the critical organ, though other
organs may also be affected. The metals differ also in regard to

their metabolism. Thus some undergo substantial changes in oxidation state or are extensively alkylated or dealkylated, depending upon the form administered, for example, mercury and arsenic. Others, by contrast, appear to be quite stable in the forms to which humans are usually exposed, for example, cadmium and lead. This broad range of characteristics will become evident in the review which follows. The metals under review were specifically selected with these aspects in mind. Selection was also based on the toxicological significance of the metals and on the depth of available knowledge. For many metals the amount of available information is relatively scanty and they are not currently considered to be of major toxicological concern. Thus, for example, both antimony and thallium are highly toxic but inadvertent human exposure at toxic levels is rare today. Concerns for the toxicity of other metals may arise in the future. Current knowledge concerning the metals reviewed here will help in evaluating their toxicological significance and could serve as a guide in their future characterization. Many new, interesting, and unpredictable phenomena will no doubt emerge.

2. LEAD

2.1. General Aspects

Lead has been recognized as a poison since the time of the ancient Greeks, as an occupational problem among miners, and later as an adulterant of wine. In modern times lead poisoning became the object of intensive toxicological investigations, first because of widespread poisoning in such occupations as enameling, smelting, and battery manufacturing and later as a result of the ingestion of lead-based paints by infants and young children. Interest in the subject was further stimulated by the realization that widespread environmental contamination has occurred as a result of the atmospheric emission of lead fumes from industrial sources and, more recently, as a result of the widespread use of lead as a gasoline additive in the form of

tetraethyl and tetramethyl lead. Although these alkyl forms of lead
have been a substantial source of environmental contamination, they
undergo combustion prior to their atmospheric dissipation, predominantly as oxides, carbonates, and sulfates. In dealing with the
numerous specific issues of lead toxicology, it is assumed that the
toxic species involved is inorganic lead.

2.2. Absorption, Distribution, and Excretion

The broad features of lead disposition in humans have been known for
many years, largely based on the pioneering investigations of lead
metabolism in humans conducted by R. A. Kehoe and his staff from
about 1940 to 1968. These studies have shown that lead administered
chronically to human volunteers, either orally or by inhalation,
accumulates in the body with no apparent attainment of steady-state
conditions over several years [1]. More recent autopsy studies suggest that while lead accumulates in bone well into middle age, no
further accumulation may occur beyond that period [2]. By contrast,
there was no perceptible increase noted in most soft tissues beyond
the second decade of life.

It has long been known from animal studies that the gastrointestinal absorption of lead is influenced by dietary conditions,
lead being reduced by calcium, phosphate, and iron and by suboptimal
vitamin D. A recent study suggests that intestinal transport of lead
is at least partly capacity limited and energy dependent [3].

All studies dealing with lead distribution within the body
indicate that in adults approximately 95% of the body burden of lead
is in bone. By contrast, only approximately 70% of the body burden
is distributed to bone in young children [4]. Thus distribution of
lead within the body under conditions of long-term intake is a slow
process, at least so far as distribution between bone and other tissues is concerned. The above observations have certain important
implications regarding interpretation of the impact of lead exposure
on the concentration of lead circulating in the blood. As an example,

the impact of this large pool of lead in bone on the variable die-
away of blood lead following the termination of periods of high lead
exposure becomes apparent. It is known from isotope tracer studies
that upon cessation of administration of a rare lead isotope for a
month or so, the $T_{1/2}$ for the return of blood lead to baseline is of
ᴛhe order of 30-40 days [5], whereas with termination of long-term
occupational exposure, the $T_{1/2}$ is of the order of several years [6].
In the first instance, clearance of lead from blood to bone probably
adds considerably more to clearance by excretion than in the second
instance, since the net transfer of lead into bone probably is much
greater than in the second instance.

The concentration of lead in the blood has long served as the
most widely accepted surrogate for current and recent biologically
available internal doses, at least in studies involving the toxicity
of lead in humans. Moreover, and as a consequence of this situation,
dose-effect and dose-response relationships concerning toxic effects
of lead have been largely developed utilizing the concentration of
lead in the blood as a measure of dose. This is generally expressed
in micrograms of lead per deciliter (µg Pb/dl) of whole blood and is
commonly designated in those terms as PbB. It is important to note
in this connection that there is no basis for assuming that a linear
relationship exists between PbB and the rate of lead intake, or
between PbB and the concentration of lead at receptor sites for lead
toxicity. Quite to the contrary, all available evidence suggests a
nonlinear relationship between PbB and lead intake, as well as between
PbB and lead at receptors for toxicity. As the concentration of lead
in air and drinking water increases linearᴌy, the corresponding incre-
mental rises in PbB become progressively smaller [7,8]. Animal
studies show that for other tissues, notably bone, a near-linear
relationship exists between oral lead intake and tissue Pb concentration. Thus there is for many tissues a disproportionately large
increase in Pb content with linear increase in PbB. That this dis-
proportionality may be reflected in the dose dependence of certain
effects of lead is shown by the nonlinearity between PbB and the
effect of lead on the hematopoietic system. It has been demonstrated

that there is a disproportionately large rise in erythrocyte aminolevulinic acid dehydratase (ALAD) inhibition [9] and in erythrocyte
protoporphyrin concentration [10,11] with increases in PbB.

Some studies suggest that the body responds to the presence of
lead by the induction of two lead-sequestering mechanisms. The first
of these involves the appearance of intranuclear inclusion bodies
which have high concentrations of lead. These have been noted to
occur with high lead exposure in both humans [12] and experimental
animals [13]. High lead exposure in occupationally exposed humans
has also been shown to result in the apparent induction of a low
molecular weight lead-binding protein in blood [14]. There was considerable variation among individuals concerning the appearance of
this lead-binding protein. Those who responded with the appearance
of this protein showed lesser evidence of lead toxicity than those
who did not, as measured by inhibition of erythrocyte adenosine
triphosphatase.

2.3. Toxicity

2.3.1. Effects on Heme and Hemoproteins

Since the initial observation by Binnendijk of excess porphyrin in
the urine of a patient with lead poisoning (as cited in Ref. 15), a
vast literature has amassed regarding the effects of lead on the
synthesis and degradation of heme and hemoproteins. This literature
has been reviewed on repeated occasions (see, e.g., Ref. 16). The
major biochemical features, observed in both humans and experimental
animals, are (1) increased erythrocyte protoporphyrin IX, (2) increased urinary and plasma aminolevulinic acid (ALA), and (3) decreased blood ALAD activity. The first effect is ascribed to inhibition of heme synthetase, the mitochondrial enzyme which catalyzes
the insertion of iron into protoporphyrin IX, forming heme, and the
second effect is attributed to inhibition of the cytosolic enzyme
ALAD, which catalyzes the conversion of ALA to the monopyrrole por-
phobilinogen. These effects begin to occur at levels of lead expo-

sure only slightly above the range which is considered normal in the
U.S. population, PbB = 10, 15, and 25 μg Pb/dl for blood ALAD inhibition, erythrocytic protoporphryin, and urinary ALA excretion, respectively, as compared to a normal PbB of approximately 8-15 μg/dl.
Reduction in circulating hemoglobin concentration, however, probably
begins to occur only at a somewhat higher PbB, though a threshold for
reduction in hemoglobin has yet to be defined with any degree of precision. The higher level of lead exposure required to depress hemoglobin concentration is probably due to the fact that lead-induced
inhibition of heme synthetase causes a compensatory derepression of
ALA synthetase, the rate-limiting enzyme involved in heme synthesis
[17].

A further effect of lead on heme has been noted which tends to
reduce circulating hemoglobin. It has been observed that lead, as
well as some other metals, enhances the biodegradation of heme by
increasing the activity of heme oxygenase [17,18]. This effect is
observed to involve not only hemoglobin but also a number of other
hemoproteins, including cytochrome P_{450} and cytochromes a, b, c, and
c_1 [18,19].

It should be noted at this point, however, that a reduction of
hemoglobin can result from an entirely different effect of lead, one
on the integrity of the erythrocyte membrane. Studies have shown
that excessive lead exposure causes a reduction of the mean life
span of circulating erythrocytes in both adults [20] and children
[21]. Animal studies suggest that this is due to oxidative damage
of the erythrocyte membrane [22,23]. The relative importance of
effects of lead on heme synthesis, heme and hemoglobin degradation,
and oxidative damage to erythrocytes as determinants of the anemic
effect of lead is not known at this time.

In consideration of the fact that lead has depressive effects
on the level of such a wide variety of biologically important hemoproteins, it is tempting to invoke a unitarian hypothesis of lead
toxicity wherein all toxic effects are attributed to a reduction in
hemoproteins. Evidence to date does not support this concept. Thus,

for example, Holtzman et al. [24] studied the effect of lead on brain
hemoproteins at levels of exposure in infant rats which compromised
growth. No effects were found on cytochromes P_{450}, $c + c_1$, $a + a_3$,
b, and b_5. Certain other effects of lead of current interest may
nonetheless involve effects on hemoproteins but have not as yet been
studied as to their mechanism of action. An outstanding example is
the recently reported depressive effect of lead on serum 1,25-di-
hydroxycholecalciferol in children [25]. This effect was observed
at PbB levels only slightly above normal. It is quite conceivable
that this effect is due to inhibition of the conversion of 25-
dihydroxycholecalciferol to the biologically active 1,25 metabolite,
since the converting enzyme is probably a cytochrome P_{450} [26].
Such a possibility is supported by studies in adults [27] and children [28] showing that P_{450}-dependent drug metabolism is retarded
in lead-exposed subjects.

2.3.2. Effects on the Peripheral and Central Nervous System

There is a voluminous literature documenting a wide range of toxic
effects of lead on both the peripheral and central nervous system.
At one time very serious crippling effects were observed in both
adults and children. Paralysis of nerves supplying flexor and
extensor muscles of the extremities was frequent (lead palsy).
Serious and often fatal derangement of the central nervous system
was characterized by convulsions and/or coma (lead encephalopathy).
These effects were often fatal or frequently irreversible. In more
recent years attention has turned to the more subtle effects of lead
on the nervous system.

   In the case of the peripheral nervous system palsy is no longer
observed, but some studies have shown that at the lower levels of
lead exposure which exist in certain occupations slowing of nerve
conduction velocity may occur in the absence of palsy [29,30]. The
mechanism is not known, but it apparently does not involve axonal
degeneration or demyelination [29].

Recent investigations concerning the effects of lead on the nervous system have been of three very distinct types: (1) electrophysiological, (2) neurochemical and (3) neurobehavioral. Only a few studies have attempted to link electrophysiological effects to neurochemical effects in the same system or neurobehavioral effects to either neurochemical or electrophysiological effects. Studies linking electrical events to neurochemical events have been largely limited to isolated nerve and nerve-muscle preparations. Both ganglionic and neuromuscular postsynaptic evoked potentials are inhibited by lead, an effect that is reversed by calcium and which is due to inhibition of acetylcholine release [31,32] or release of norepinephrine at sympathetic nerve terminals [33].

Electroencephalographic studies conducted recently in children with only moderately elevated lead exposure and no overt signs of toxicity indicate that lead has a dose-related effect on slow-wave voltage, an interaction noted even at PbB < 30 μg/dl [34]. Animal studies tend to support the implication that very moderate degrees of lead exposure cause alterations in cortical function, such as excitability of the visual cortex [35]. Neither neurochemical nor neurobehavioral correlates of these findings have been reported. Thus their human health significance is uncertain.

The effects of lead on neurotransmitter systems in the brain have been the subject of numerous animal studies. These have been reviewed recently [36]. As a result of the link noted between electrophysiological and neurotransmitter function in peripheral nerves, most of these studies have focused on cholinergic, mono-aminergic (dopamine, norepinephrine, 5-hydroxytryptamine), and GABAergic (gamma-aminobutyric acid) systems. The role of calcium has been included in many of these studies, since it is involved in transmitter release, synthesis, and storage. The mechanisms whereby altered transmitter function may account for behavioral and electrophysiological effects remain elusive, however. This is in part due to the fact that no specific site of action in the brain has been identified.

In recent years studies concerning the neurobehavioral effects
of lead have focused mainly on subtle effects occurring as a result
of moderately elevated exposure in young children. The major problems
with the studies reported in children have been (1) a lack of adequate
accounting for covariates known to influence the cognitive or behavioral variables measured and (2) uncertainties regarding the actual
levels of lead exposure of the subjects prior to testing for lead
effects. This literature has recently been critically reviewed [37].
While the occurrence of subtle neurobehavioral effects seems plausible, the level, duration, and age at which the required insult occurs
remain to be determined. For that matter, even the precise nature of
the effect(s) is still uncertain.

A vast literature exists concerning effects of lead on behavior
using animal models and paradigms originating from the field of experimental psychology. Much of this work is flawed in the same manner as
the human studies: through lack of adequate consideration of other
variables influencing behavioral outcome and poor definition of the
internal dose of the lead. This literature has recently been critically reviewed with particular attention to interpretive difficulties
regarding human health implications [38]. When most of the usual
flaws in experimental design are eliminated, effects of lead are
often no longer apparent [39], and there always remains the question
whether the functions being studied are relevant to the human condition.

2.3.3. Other Effects

The toxic effects of lead are by no means limited to the hematopoietic
and nervous systems. Thus lead has long been known to be nephrotoxic,
to affect the control of blood pressure via the renin-angiotensin
system, and to affect both the male and female reproductive system.
It is not clear at this time, however, whether these effects are
significant at the prevailing levels of human exposure. The U.S.
Environmental Protection Agency is preparing a comprehensive air lead
criteria document which cites the most recent published information

concerning lead toxicity. This document should be available for
distribution by mid-1985.

3. ARSENIC

3.1. General Aspects

Arsenic, often addressed as half-metal, has long been the subject of
toxicological research. Early studies resulted from its extensive
use for homicidal and suicidal purposes, as a pesticide in orchards
and vineyards, and as a vesicant in chemical warfare. Arsenic is
still extensively used as a herbicide in agriculture, as a silvicide
in forestry, and as a wood preservative. It is also finding increasing use as a substrate material in the integrated circuit industry,
replacing silicon-based technology.

   Human exposure is obviously associated with the above uses.
Significant human exposure also results from the smelting and refining
of metallic ores, particularly copper ores, and in association with
the combustion of coal.

   Unlike lead, arsenic occurs in the environment in a wide variety of relatively stable chemical forms. Its two primary oxidation
states are As(III) and As(V). In nature the primary inorganic forms
are pentavalent, though trivalent compounds are readily found in
water, particularly under anaerobic conditions. Some methyl arsenic
compounds such as dimethylarsinic acid are also found in nature as a
result of microbiological activity. Dimethylarsinic acid and methyl-
arsonic acid are also manufactured for use as selective herbicides
in agriculture and forestry.

3.2. Absorption, Distribution, and Excretion

Arsenic in most chemical forms is readily absorbed by the human
gastrointestinal tract. This is based on studies of the urinary
excretion of single oral doses of inorganic As(V) and As(III),

[40-42], arsenic in seafood [42], and methylated arsenic [43]. These
diverse forms are all excreted to the extent of 46% or more of the
dose within a few days, with the methylated forms being excreted to
a higher degree than the inorganic forms. Urinary excretion of
orally administered doses provides only a minimal estimate of absorption, since it does not take into account arsenic which may have been
absorbed and subsequently excreted by other routes, notably the gastrointestinal tract. The one exception concerning absorption of
arsenic in humans is the poorly soluble selenide As_2Se_3. Oral doses
of 12 mg resulted in no increased arsenic excretion in the urine
[43]. The very substantial absorption of orally administered arsenic
in human has also been noted in a variety of animal species, including
the rat, pigs, and monkeys.

There are no studies concerning pulmonary absorption of arsenic
in humans. Studies of arsenic retention following intratracheal
instillation in rats [44] and hamsters [45] indicate that the retention of calcium arsenate is much greater than the retention of either
arsenic trioxide or arsenic trisulfide.

Once absorbed, arsenic undergoes substantial metabolism, as
reflected in the patterns of arsenic compounds found in the urine.
Knowledge regarding the biotransformation of arsenic in the body is
only fairly recent. In humans, and in most animal species for that
matter, both inorganic As(III) and As(V) administered orally appear
in the urine predominantly as dimethylarsinic acid (DMA) and, to a
lesser degree, as monomethylarsinic acid (MMA) [42]. The sum of
DMA, MMA, and inorganic arsenic in the urine accounts for essentially
all the arsenic administered [46]. So far as the pathway of metabolism is concerned, it appears that DMA is the terminal metabolite on
the basis that, when administered orally to humans, it is excreted
in the urine unchanged, while there is some slight conversion of
orally administered MMA to DMA. Based upon studies in rabbits, it
appears that the biomethylation of inorganic arsenic requires the
initial reduction of As(V) to As(III), an event which precedes the
appearance of DMA in the urine [47]. The metabolic pathway from

As(V) to DMA appears to be efficient within reasonable limits of
dosage in humans, up to 500 μg of As. At much higher, near-suicidal
doses, however, the emergence of DMA as the dominant urinary metabolite seems to be somewhat delayed [42].

The distribution of arsenic in the body has been studied in
both humans and animals. Under conditions of chronic intake, the
highest concentrations are found in hair, nails, and skin. These
all have high sulfhydryl concentrations, suggesting that this prop-
erty accounts for their high affinity for arsenic. Based on human
autopsy studies, it does not appear that arsenic is highly accumu-
lative over the long term. Pharmacokinetic studies support this
conclusion. The biological half-life of As(III) administered orally
to humans is 39-59 hr, depending upon the dose, and with daily administration the steady-state urinary excretion rate is attained within
5-6 days [43,46].

3.3. Toxicity

The toxic effects of arsenic as described in humans vary in accordance
with the dose and the duration of intake. Thus acute arsenic poisoning resulting from large doses taken over a period of one or several
days differs from the disease which occurs when much smaller doses
are taken over months to several years. In the case of acute poisoning the predominant effects are severe gastrointestinal irritation
with attendant cramps and diarrhea, effects which have an almost immediate onset. This is usually followed by the onset of peripheral
neuropathy 1 to several weeks later. Both sensory and motor function
of peripheral nerves are affected in a manner common to a number of
different toxic substances. The process is one of "dying back,"
wherein the most peripheral regions of the axons are first affected.
It has been postulated that axons are peculiarly sensitive to poisons,
as compared to their neurons, because they are not capable of gener-

ating cofactors essential for energy generation; rather, they are dependent upon the neuronal cell body [48]. Axonal regeneration often occurs with a slow return to normal function. The precise mechanism whereby axonal degeneration occurs is not known. No animal model for this disease has been reported, although chickens fed high levels of arsanilic acid develop paralysis of the limbs [49].

In a detailed description of 41 cases of arsenic poisoning certain other manifestations were noted, though less frequently than peripheral neuropathy [50]. These included hyperkeratosis, scaling and pigmentation of the skin, drowsiness, poor memory, and transient episodes of irrational speech and delirium. Anemia and leukopenia were also noted in some cases.

In another series of cases, attributed to the use of arsenic-contaminated herbal medicines, the same signs and symptoms were noted, though skin lesions were most frequently noted [51]. This may have related to the more chronic nature of intake.

Not much is known concerning the mechanisms whereby nonneurological toxic effects arise. Older studies have shown that As(III) reacts primarily with thiols in tissues, whereas As(V) interferes with PO_4 metabolism and oxidative phosphorylation. The anemia which is frequently observed may be due to inhibition of two enzymes involved in heme synthesis, aminolevulinic acid synthetase and heme synthetase, effects noted in rats receiving sodium arsenate in their diet for periods up to 6 weeks [52].

Arsenic is a carcinogen in humans. Numerous epidemiological studies have indicated a strong positive association between atmospheric levels of arsenic in copper smelters and the incidence of lung cancer. Skin cancer also is associated with long-term arsenic exposure by dermal contact and by oral intake. Cancer due to arsenic is reviewed in another chapter.

4. VANADIUM

4.1. General Aspects

Down through the years attention has been directed mainly to metals which have gained prominence as a result of well-publicized, frequent occurrences of poisoning, notably with lead, cadmium, arsenic, and mercury. Many other metals are no doubt equally toxic but have received lesser attention, perhaps because human cases of poisoning have been reported less frequently or have been less dramatic. Vanadium clearly falls in this category. Only a limited number of reports have been published concerning human toxicity. It has been selected for review because it has rather interesting toxicological properties and because its production is forecast to increase [53].

The major use of vanadium is as an alloying element, mainly in the steel industry. The demand for and production of vanadium have risen steadily in the United States over the past 25 years and are likely to continue to increase. Vanadium exists in valence states of +2, +3, +4, and +5, the last two being the most stable. It is found in nature as oxyanions that tend to combine with ligands, for example, phosphorus and sulfur.

Human exposure is greatest among workers engaged in the extraction of vanadium from ores or as a result of contact with the residue of combusted petroleum in boilers. These residues are sufficiently rich in vanadium to have been used as a commercial source of vanadium.

4.2. Absorption, Distribution, and Excretion

The oral absorption of vanadium in humans can be estimated from only one study [54]. Subjects received ammonium vanadyl tartrate [$(NH_4)_4(VO)_2(C_4H_2O_6)$ · $2H_2O$, 25-125 mg/day] for 45-68 days. On the average, 2.4% of the dose was excreted in the urine. In rats, 31%

of single oral doses of sodium vanadate was absorbed, based on the
fact that 69% was excreted in the feces [55].

The highest concentration of vanadium found in human tissues
at autopsy is in the lungs: 19-140 ng/g wet weight as compared with
less than 7 ng/g in other organs [56]. This selective concentration
in the lungs has not been observed in animals receiving vanadium
orally or by parenteral injection. The valence of the vanadium
administered does not seem to influence the pattern of organ distribution, at least not in rats [57]. This lack of differential behavior based on valence of the administered compound is probably due to
the fact that V(V) is quickly reduced to V(IV) in the body [58,59].

Nothing is known concerning the pharmacokinetic aspects of
vanadium distribution and excretion in humans. The only age-related
increase reported has been in lungs [60].

The primary route of vanadium excretion in humans is the urine.
This is based on the recovery of intravenously administered sodium
tetravanadate in human volunteers [61]. The kidney also is the major
route of excretion in animals [62,63].

4.3. Toxicity

The only naturally occurring cases of vanadium poisoning in humans
have resulted from the inhalation of vanadium fumes. The effects
are largely limited to the irritative actions on the respiratory
epithelium. Based on the finding of bronchial hyperreactivity to
histamine, vanadium may also act as an asthmogen [64].

When taken by the oral route, vanadium causes gastrointestinal
disturbances in the absence of any other clearly defined manifestations of toxicity. These effects have been documented by several
groups of investigators who administered vanadium orally to human
volunteers in the course of studies concerning the possible hypo-
cholesterolemic effects of vanadium [54,65,66].

Interest in the toxicological properties of vanadium was stimulated in 1977 by a report indicating that it is a potent inhibitor

of Na^+,K^+-ATPase [67]. The analogy to ouabain was obvious, leading
to numerous studies of the cardiovascular and renal effects of vanadium, since ouabain has major pharmacological actions on these two
organs/systems.

Vanadium has been shown in numerous studies to be a vasocon-
strictor. The role of Na^+,K^+-ATPase inhibition is uncertain. On
the one hand, the vasoconstrictive effect of vanadium on isolated
coronary vessels is attenuated or eliminated by ouabain, suggesting
mediation by inhibition of Na^+,K^+-ATPase [68]. On the other hand,
the increased tension developed by canine saphenous vein preparations
does not appear to be due to inhibition of Na^+,K^+ pump activity [69].
Although, like ouabain, vanadium has been shown to have a positive
inotropic effect on cardiac tissue, a causal link to Na^+,K^+-ATPase
inhibition, is questionable [70]. Doubt also has been cast on the
role of vanadium-induced inhibition on Na^+,K^+-ATPase in its ouabain-
like action on intestinal smooth muscle [71].

Vanadium has also been shown to have a peroxidative effect on
hepatocytes, like a number of other metals. Although vanadium was
the most potent in regard to lipid peroxidation, it was the least
potent in terms of cellular injury [72]. This suggests that cellular
injury may not be due to the peroxidative effect of vanadium.

It seems from all the above studies that much remains to be
learned concerning the mechanisms of vanadium toxicity. Some have
suggested that it may have an essential function in the regulation
of smooth muscle contraction.

5. MERCURY

5.1. General Aspects

Mercury poisoning has long been recognized in humans, resulting from
both occupational and nonoccupational exposure. Excessive occupational exposure is due mainly to inhalation of mercury vapor in the
electrolytic production of chlorine and sodium hydroxide and in the

manufacture of numerous items requiring elemental mercury, for example, thermometers. In the general population there have been major epidemics of mercury poisoning due to methyl mercury, resulting either from the use of mercury-contaminated seed grain for the production of flour or from the contamination of fish from mercury in industrial effluents.

It is partly as a result of these epidemics that the production and use of mercury have been reduced in recent years. Worldwide production has fallen from 10,364 metric tons in 1971 to 6051 in 1978 [73]. The agricultural uses of mercury have been largely curtailed in the United States, resulting in a reduction in these uses from 118 metric tons in 1968 to 21 metric tons in 1975 [73].

Mercury exists in nature as the metallic vapor (Hg^0) by volatilization from the earth's land mass, as inorganic salts of Hg(I) and Hg(II), and as monomethyl and dimethyl mercury ($MeHg^+$, Me_2Hg). Monomethyl mercury is found in organic sediments from both fresh water and coastal waters. The methylation process results from the action of microorganisms. From the point of view of human exposure, the methylation process is extremely significant, since this form bioaccumulates in the aquatic food chain, resulting in high concentrations in fish marketed for human consumption [74].

5.2. Absorption, Distribution, and Excretion

Mercury as Hg^0 is poorly absorbed from the gastrointestinal tract (<1% based on animal studies) but readily absorbed following inhalation of Hg^0 vapor. Gastrointestinal absorption of ^{203}Hg-labeled mercury has been studied in humans. The absorption of tracer doses of inorganic mercury taken with food is approximately 7% [75]. By contrast, 95% of $MeHg^+$ is absorbed, regardless of whether taken with food or not [76]. Young rats (7 and 23 days old) absorb Hg^{2+} much more efficiently than mature rats (7 weeks). Two factors are involved. The first is higher intestinal transport in young rats as determined by utilizing closed intestinal segments. The second is the slower gastrointestinal transit time in the young rat [77]. Gastrointestinal absorption of MeHg$^+$ has been shown to be depressed by the normal intestinal microflora in the rat [78]. This is in keeping with the observation [78] that intestinal bacteria from both the rat and man can demethylate MeHg$^+$.

The pattern of distribution of mercury in the body is very much dependent upon the chemical form: Hg^0, Hg^{2+} and MeHg$^+$ all have highest affinity for the kidney, but inhaled Hg^0 and orally administered MeHg$^+$ attain higher concentrations in the central nervous system as compared to Hg^{2+}. This difference is attributed to the higher lipid solubility of Hg^0 and MeHg$^+$ than that of Hg^{2+}. The differences are sufficient to influence materially the nature of the toxicity. The distribution of phenyl mercury is essentially the same as for Hg^{2+} owing to rapid cleavage of the carbon-mercury bond [79]. Cleavage of the carbon-mercury bond also occurs in the case of MeHg$^+$, although more slowly and at various rates, depending upon the organ. At least this is so in rats [80]. Cleavage is much lower in the brain than in the liver or kidney. Also Hg^0 undergoes substantial metabolism, being oxidized intracellularly to Hg^{2+}.

The major pathway of mercury excretion varies in accordance with the form administered. In the case of MeHg$^+$, fecal excretion substantially exceeds urinary excretion. About half appears in the feces as inorganic mercury. By contrast, only approximately 10% of the urinary mercury excreted following MeHg$^+$ is inorganic [80]. In the case of Hg^{2+} the predominant route of excretion is the urine.

There is limited information concerning the pharmacokinetics of MeHg$^+$ and Hg^{2+} in humans. It has been shown that MeHg$^+$ is cleared from the body with a half-time of 76 days, while the half-time was 42 days following administration of Hg^{2+} [76]. The longer half-time for MeHg$^+$ is somewhat deceptive in the sense that it is based on total mercury in the body, not taking into account that a substantial fraction of the body burden undergoes demethylation, at least based on studies in rats [80]. True half-time for MeHg$^+$ would thus be shorter to some degree if this metabolized fraction were taken into account.

5.3. Toxicity

The toxic effects of mercury depend upon the form administered. The kidneys are extremely sensitive to mercury, largely because a major portion of the body burden accumulates in this organ. Certain organic derivatives of mercury have been used as diuretics on this basis. Inorganic salts of mercury (Hg^{2+}) cause renal damage, while Hg^0 and $MeHg^+$ cause mainly disturbances of the peripheral and central nervous system. This particular distinction is readily explainable on the basis of differential body distribution. Fractional distribution of Hg^{2+} to the brain is far smaller than for Hg^0 or $MeHg^+$. The most controversial current issue concerns the distinction between the effects of Hg^0 and $MeHg^+$ and why these differences exist. The effects of mercury vapor exposure are neuropsychiatric in nature, whereas the effects of $MeHg^+$ are largely sensorimotor. Both forms cause intention tremor. The neuropsychiatric complex observed in mercury vapor (Hg^0) exposure consists of excessive shyness, insomnia, and emotional instability. This complex is referred to as erethism.

While tremor is frequently reported in cases of $MeHg^+$ poisoning, sensorimotor disturbances not seen in mercury vapor poisoning are very frequent, for example, ataxia, impaired gait, constriction of the visual field, and disturbance of swallowing [81]. These dissimilarities have led to the conclusion that the neurotoxicities of $MeHg^+$ and of Hg^0 are due to these two distinct chemical species. An alternate explanation for the differential nature of these two diseases is that regional distributions in the brain are different [81], thereby affecting different neuronal areas. The two forms of mercury certainly produce the same type of neuropathological effects on neuronal elements [82]. It has been argued that the toxic effects of $MeHg^+$ on the brain must be due to $MeHg^+$ per se rather than to inorganic mercury resulting from demethylation because so little demethylation occurs in the brain [80]. This has recently been disputed on the basis that Hg^{2+} is 100-fold more potent than $MeHg^+$ as an inhibitor of brain muscarinic receptors [83]. No other studies have compared the

potency of these two forms of mercury on other discrete quantifiable
receptor-agonist interactions.

Some advances have recently been reported concerning the mechanism whereby Hg^{2+} exerts its toxic effects on nonneuronal tissue.
It has been shown that Hg^{2+} induces lipid peroxidation in isolated
rat hepatocytes in a dose-related manner [84]. This effect was related to the degree of lactate dehydrogenase release from the cells
(an expression of cellular toxicity). This peroxidative effect was
probably not the cause of cellular toxicity, however, since two anti-
oxidants inhibited peroxidation but did not alter lactate dehydrogenase release. In another study the mechanism of the vascular
toxicity of Hg^{2+} was examined [85]. It was found that mercury-
induced elevation in baseline vascular pressure could be prevented
by lowering the extracellular calcium concentration and by introducing the calcium channel blocker lanthanum into the perfusing
medium. This is consistent with the hypothesis that the entry of
calcium into cells is the common pathway for cell death regardless
of the inciting agent [86].

6. CADMIUM

6.1. General Aspects

The element Cd exists in various forms throughout the lithosphere.
Of special relevance to environmental concerns is its presence in
mixed Zn ores, fossil fuels, and phosphate rock, the main constituent
of superphosphate fertilizers. Two properties of Cd lead to important
environmental consequences: (1) the relatively high vapor pressure,
which accounts for its ready vaporization, for example, during smelting operations or coal combustion, and (2) the high solubility in
water, especially at a slightly acid pH.

The wide distribution of Cd in fuels, fertilizers, and mine
tailings, together with its extensive use in nonrecycled form in
industry, accounts for a gradually increasing concentration of the

metal in the environment. Low natural levels may be raised significantly, especially near industrial point sources such as smelting
operations. As a result, not only groups of industrial workers but
also the general population have increasingly been exposed to Cd,
and Cd toxicity has become a major problem of environmental pollution. The so-called itai-itai disease in Japan, characterized by
osteomalacia and renal tubular malfunction, has been attributed to
Cd in irrigation water. The great concern with Cd is reflected in
the appearance of several critical review volumes (e.g., Refs. 87
and 88), which should be consulted for more detailed discussion and
documentation than can be offered here.

There is little evidence to suggest that Cd is an essential
element in mammalian nutrition. This conclusion is reinforced by
the observation that, at least in the mouse, the rate of Cd uptake
is independent of any homeostatic control [89]. Instead, uptake
continues independently of the body burden of metal. Excretion,
moreover, is slow, so that cadmium accumulates during chronic
exposure.

6.2. Absorption, Distribution, and Excretion

Outside of the industrial environment the body burden of Cd in
animals and humans is derived mainly from food and water. An additional and significant source of body Cd may be tobacco smoke;
inhalation exposure, where it occurs, is especially significant,
not only because it may lead to fibrotic changes in the lungs but
also because fractional absorption of inhaled Cd is much higher
than that of the ingested metal.

Like other heavy metals, Cd in food does not readily cross the
intestinal barrier. Values reported for fractional absorption from
the intestine mostly range from 1 to 5%. Most experimentally determined values for the fractional uptake of Cd actually lead to values
of net uptake or retention, ignoring the occurrence of biliary or
enteric secretion of Cd. For instance, the demonstration that

retention of Cd in rats on a milk diet significantly exceeds that
with other kinds of diet [90] provides only limited information on
the unidirectional flux of Cd out of the intestinal lumen.

A great many exogenous and endogenous factors help determine
Cd absorption. These include physiological variables such as age,
sex, lactation, and the composition of the diet. Polyvalent cations
such as Ca, Zn, and La depress Cd absorption, apparently by altering
the ability of Cd to bind to the brush border membranes [91]. A
major fraction of the Cd removed from the intestinal lumen remains
trapped in the intestinal mucosa and is lost to the body upon desqua-
mation of the epithelial cells. The nature of this trapping has not
yet been fully elucidated.

The result of all these variables is that the magnitude of Cd
absorption may vary between wide limits. In general, however, and
excluding the neonatal period during which metals cross the intestinal barrier relatively readily, higher animals and humans absorb
only a very small fraction of ingested Cd, somewhere between 1 and
5%. The biological half-life of Cd in the body is very long, however, of the order of 20 years, so that Cd acts as a cumulative
poison. Measurement of total body burden in vivo has recently
become possible with the aid of neutron activation techniques [92].

The distribution of Cd in the body depends on the form in
which it reaches the blood. Inorganic Cd tends to accumulate first
in the liver; it is also taken up by other organs such as the testes.
Cadmium administered as thiol complex can be more readily taken up
by the kidneys, as further described in Sec. 6.3. A small amount of
Cd only continues to circulate in blood, and the normal level of
around 1 μg/dl is readily raised during Cd exposure. The circulating
Cd may be bound in part to metallothionein (see Sec. 6.3), a filterable low molecular weight protein readily taken up by the kidney.
The major portion of total body Cd following prolonged exposure is
accumulated in the kidney cortex. As a result, kidney levels of the
metal increase with age. The ultimate result is the production of
renal lesions; at that stage the accumulated Cd may be released from
the kidney and gradually excreted in urine.

The major route of excretion of Cd is via the urine, although
when expressed as a percentage of the total body content of Cd, the
excretion is very small. Normal values fall into the range of 1-2
µg/day in human male adults on a North American diet. Increased
urinary levels are noted in smokers or as a result of any increased
Cd exposure. When the kidneys have been damaged as a result of long-
term exposure, they may release their accumulated Cd and a sharp rise
in Cd excretion is seen.

6.3. Toxicity

Like other heavy metals, cadmium reacts readily with proteins and
other biological molecules. The chemical similarity of Cd to Zn
suggests the hypothesis that in many cases such reactions may dis-
place Zn, for instance, from a critical site in Zn enzymes. It is
not surprising, then, that Cd exposure may result in effects at
multiple sites, including the lungs, prostate, testes, heart, liver,
and kidneys; in chronic exposure the primary target organ is the
kidney. In line with the observations on the distribution of acutely
administered $CdCl_2$ and Cd-thiol compounds in the body, the kidney
does not take up Cd after acute exposure to inorganic compounds:
Under these conditions the testes are peculiarly sensitive to the
metal.

Cardiovascular effects of cadmium have been repeatedly
described. Several investigators reported the experimental production of hypertension in rats chronically exposed to low concentrations of Cd [93]. Whether this observation has much bearing on
human essential hypertension is uncertain. Thus the inhabitants of
Cd-polluted areas in Japan do not show an increased incidence of
this disease.

After long-term chronic exposure, the kidneys accumulate a
major portion of total body Cd. The renal cortex is also the major
target organ of Cd toxicity. The symptoms produced resemble those
of the Fanconi syndrome, with inhibition of various proximal tubular

transport mechanisms. In distinction from the tubular lesions seen
in poisoning with other metals, cadmium especially interferes with
the tubular reabsorption of low molecular weight proteins. Excretion
of β_2-microglobulin is one of the diagnostic features of Cd intoxication.

The signs of renal malfunction appear only slowly, long after
increased excretion of Cd first reflects exposure to Cd. This fact
seems to be related to the need for renal cortical Cd levels to reach
a critical value before renal damage ensues. This concept of a critical concentration of Cd has been widely discussed. Tentatively, the
best estimate of the population critical concentration at which 10%
of the exposed individuals are likely to show signs of malfunction
(PCC-10) has been set near 200 μg/g cortex (fresh weight) [94].
While the evidence in support of this value for the human populations
studied is quite strong, it may not necessarily be applicable to
other populations studied under different conditions. Factors such
as ambient temperature may change the critical concentration. Another important variable is the rate of Cd exposure; in acutely exposed rats low molecular weight proteinuria was seen at Cd levels as
low as 10 μg/g.

One of the major reasons for the differential sensitivity of
various populations is probably related to the effect of different
exposure conditions on Cd metabolism. In the body cadmium and other
metals induce synthesis of metallothionein (MT), a low molecular
weight protein with a high affinity for Cd [95]. The precise function, if any, of MT in Cd metabolism has not yet been elucidated.
Binding of Cd to MT in the liver, kidneys, and other organs may
represent an incidental extension of the storage role of the protein
for essential metals like Zn. The chelation of Cd by intracellular
MT interferes with the efficacy of conventional chelating drugs like
ethylenediamine-N,N,N',N'-tetraacetate (EDTA) in removing Cd from
the body. Metallothionein may be involved in the transport of Cd
from the liver to the kidney [96].

The suggested detoxifying function of MT must be reconciled
with the further observation that injected CdMT is more nephrotoxic

than $CdCl_2$ [97]. This apparent discrepancy can be explained by the
fact that any chelator possessing a sufficiently high affinity for
Cd to compete for it effectively with nonfilterable plasma protein
will permit plasma Cd to remain filterable; to the extent that the
filtered Cd complex can be reabsorbed in the renal tubule, the renal
toxicity of Cd will be increased [98]. A specific mechanism for the
absorption of anionic proteins like MT has been described in the
rabbit kidney [99]. Like reabsorbed proteins in general, MT is
digested by lysosomes. The Cd set free may be the form in which the
metal exerts its toxic effects; it will also induce the synthesis
of endogenous MT.

There is evidence in several systems for the distribution of
cellular Cd between more than one pool or species. In the renal
cortex these compartments can be distinguished by the rate of their
isotopic equilibration with plasma Cd [100]. The extensively studied
Cd resistance of selected mammalian cells may be related to the rapidity of their response to Cd exposure in initiating MT synthesis
[101]; the concentrations of non-MT Cd and of nuclear Cd may be the
determinants of toxicity. In any case, the existence of such compartments, in all probability varying in their toxic properties,
raises some question about the existence of a critical level of
total Cd in the renal cortex applicable to all populations. The
concept of a critical level must be applied with care in the study
of different populations under different exposure conditions.

7. NICKEL

7.1. General Aspects

Nickel is a fairly common element, widely used in metallurgy and in
chemical industry. In addition to mostly divalent salts, it also
forms organic complexes such as Ni-tetracarbonyl and Ni-dimethyl-
glyoxime. The element appears to be essential for normal growth in
animals [102], although its precise biological role has not yet been

clarified. The demonstration that the enzyme urease contains Ni
[103] suggests a possible role of this metal in ruminant N metabolism. Average dietary Ni intake more than suffices to meet the
probable requirements in humans or animals, and it is only at very
much higher concentrations that Ni compounds become overtly toxic.

7.2. Absorption, Distribution, and Excretion

While inhalation exposure to Ni compounds may be a major problem in
the occupational environment, the main source of body Ni in humans
and animals is that provided by food and water. This is true in
spite of the fact that, depending on the nature of the diet and
perhaps other factors, the fractional absorption of ingested Ni is
very low. Of the daily human intake of 0.3-0.6 mg, over 99% is
usually excreted in the feces. The mechanism responsible for removal
of Ni from the intestinal lumen resembles that which appears able to
account for the uptake of polyvalent trace metals in general [104].
Unlike Cd, however, Ni neither induces the synthesis of the metal-
binding protein metallothionein nor reacts with it in vivo [105].
This fact may explain why pretreatment with Zn does not protect
animals against toxic effects of Ni [106] and why the intestinal
mucosa retains little of the Ni taken up from the lumen, in strong
contrast with the retention of Cd. There is not much evidence for
the existence of a specific mechanism mediating Ni absorption, even
though the essentiality of a metal is often reflected in the specific
and homeostatic control of its absorption.

The normal levels of Ni circulating in plasma are low. Sunderman [107] reported a mean of 2.6 ± 0.9 µg/liter of serum in humans;
most of this Ni is bound to the albumin fraction. Environmental
exposure to Ni raises its concentration in serum and urine. Intravenously injected Ni is cleared from the blood according to a two-
compartment model [108], and the metal becomes widely distributed
in many tissues. The highest accumulation is found in the kidneys
(Table 1).

TABLE 1

Distribution of ^{63}Ni in Mice
(cpm/mg Fresh Weight)

	Hours after intraperitoneal injection	
	2	24
Lung	1,073	228
Kidney	13,149	3,634
Spleen	423	56
Liver	356	31
Heart	434	34
Serum	2,944	132

Source: Based on Ref. 118.

        The decline of renal concentration with time suggests that at
least a fraction of the renal load is transient and related to the
process of Ni excretion.   Indeed, the major route of excretion of
systemic Ni is in urine.   A portion of plasma Ni appears to circulate as filterable complexes with cysteine and other amino acids.
The mechanism of tubular reabsorption of filtered Ni and the possibility of tubular Ni secretion have not been explored in any detail.
It may be mentioned here that Ni concentrations in sweat approximate
those in urine, so that under conditions of profuse sweating significant amounts of the metal may be lost through the skin.

7.3. Toxicity

Not much is known about the specific molecular mechanisms underlying
the toxic effects of Ni.   Like other heavy metals, it is very reactive
and readily interacts with many biological molecules, including proteins and DNA [109].   Nickel ions do not bind appreciably to metallothionein but do form complexes with various amino acids.   Complex

formation with triethylenetetramine and penicillamine provides a
useful basis for chelate therapy.

Because of its low fractional absorption from the intestine,
ingested Ni is not very toxic; values of the acute LD_{50} in excess
of 100 mg/kg body weight have been reported in rats and mice [110].
This may be contrasted with an intraperitoneal LD_{50} of only approximately 10 mg/kg. Prolonged feeding of rats with a diet providing
about 50 mg Ni/kg body weight per day exerted widespread effects on
growth and metabolism [111]. Other workers observed lesser adverse
effects of relatively high levels of Ni in the diet [112].

Nickel interacts with many organ systems. It may damage the
lungs or alter cardiac contractility. However, for exposure other
than by inhalation, it accumulates primarily in the kidney and at
higher levels is clearly nephrotoxic [113]. The similarities between
the renal functional lesions caused by heavy metals and those characteristic of the generalized Fanconi syndrome have often been noted:
In both cases proximal tubular transport of different solutes is
greatly depressed. However, like the other metals, Ni in relatively
low doses is selective in the damage it induces. For instance, a
dose of $NiCl_2$ in the rabbit, sufficient to severely depress reabsorption of filtered aspartate, exerts little effect on the tubular transport of cycloleucine or glucose [114]; at higher doses the metal
causes complete renal shutdown.

The respiratory system can also become a target organ in Ni
exposure, especially in industrial settings. Such exposures may
involve the highly toxic and volatile $Ni(CO)_4$, readily taken up and
excreted by the lungs and causing pneumonitis and other pulmonary
effects. A variety of effects of aerosols of $NiCl_2$ or NiO have also
been noted. These include chronic rhinitis and sinusitis, interference with normal pulmonary defense mechanisms such as macrophage
function, and cancers of the nasal cavity and of the lung.

Depending on the chemical form and route of administration,
Ni compounds may be carcinogenic [115]. Epidemiological studies
have demonstrated an increased risk of cancers of the respiratory
tract among workers in Ni refineries [107]. No increased cancer

incidence was reported in a series of studies on experimental
animals exposed to Ni in food and water.

Nickel and its compounds are also highly allergenic. Allergy
may be produced by internal exposure to the metal in orthopedic
prostheses, cardiac pacemaker electrodes, steel sutures, and so on,
by dermal contact with Ni-containing jewelery, and by ingestion or
inhalation or contact with Ni salts. Thus up to 10% of a group of
patients seen by dermatologists exhibited positive skin reactions
to Ni [116], and ingestion of Ni may aggravate Ni dermatitis [117].

8. CHROMIUM

8.1. General Aspects

Chromium occurs in the environment mainly as the element Cr or as
compounds of Cr(III), the trivalent species, or of hexavalent Cr.
The most important commercial chromium ore is chromite, which contains
up to 38% of the element. Compounds of Cr are used very widely in
industry, with the U.S. consumption amounting to 327,000 tons in 1976.
Major industrial uses include the production of alloys, metal finishing
and corrosion control, application as mordants in the textile industry,
as constituents of paints, leather tanning, etc.

Trivalent and hexavalent Cr are readily interconvertible in
the environment. In biological systems Cr(VI) is readily reduced
to Cr(III), but the redox potential does not favor the reverse reaction. Compounds of Cr(VI) are generally more soluble in water than
are those of Cr(III). In solution, Cr(VI) usually exists not as
cation but in complex anions such as chromate and dichromate. By
contrast, Cr(III) usually remains cationic; it tends to form coordination complexes with a coordination number of 6, and the resulting
compounds or complexes are relatively insoluble in water. In natural
water supplies, under oxidizing conditions, and especially during
water treatment with oxidizing agents such as chlorine or ozone, the
relatively less toxic Cr(III) is oxidized to the more toxic derivatives of Cr(VI). This fact provides the basis for arguments against

the setting of separate permissible concentrations for the two
valence states in drinking water.

There is considerable evidence to indicate an essential role
for Cr in the metabolism of higher animals [119]. For instance, the
impaired carbohydrate metabolism seen in Cr-deficient rats or humans
can be corrected by administration of small amounts of the metal.
An organic derivative of Cr(III) has been identified and partially
characterized as the glucose tolerance factor believed to be required
for normal disposition of a glucose load. Normal Cr intake on a
North American diet barely suffices to meet the suspected essential
nutritional needs for Cr. However, adequate intake levels for Cr
(100 µg/day) fall orders of magnitude below those which even upon
prolonged exposure have been observed to cause toxic effects in
humans or animals.

8.2. Absorption, Distribution, and Excretion

Outside of the occupational environment, almost all of the body
burden of Cr is derived from ingested Cr. As with other metals, the
fractional absorption of ingested Cr compounds critically depends on
many factors and especially on the nature of the diet. It is not
possible, therefore, to assign a definite value to Cr absorption
from both food and water or from different kinds of food. The
greatly differing properties of compounds of Cr(III) and Cr(VI) and
the ability of gastric and enteric juices to reduce the one to the
other further complicate attempts to provide a quantitative description of Cr absorption.

In general, Cr(VI) is more readily absorbed than Cr(III).
There is evidence to suggest that the absorption of anionic compounds of Cr(VI) resembles that of anions in general. Direct evidence for facilitation of chromate translocation by anion transport
mechanisms is found in the work of Alexander et al. [120], who concluded that dicarboxylate and phosphate carriers mediate chromate
accumulation by rat liver mitochondria. Another indication of the

ability of Cr(VI) to penetrate cells due to its anionic nature is
its uptake across the generally anion-permeable erythrocyte membrane.
This is followed by the intracellular reduction and trapping of Cr
and forms the basis for the well-established use of chromate-51 to
label erythrocytes and other cells. Similar processes presumably
explain the greater mutagenicity of Cr(VI) than that of Cr(III) [121].
Finally, an important factor favoring absorption of Cr(VI) over that
of Cr(III) is the generally higher solubility of the hexavalent compounds at physiological pH.

Fractional absorption of inorganic Cr(III) from the intestine
is much smaller than that of Cr(VI). The common use of ^{51}Cr(III) as
a fecal marker is based upon the essentially complete fecal excretion
of the isotope when administered orally. However, some uptake of
inorganic Cr(III) is likely to occur; after all, even such nonessential and toxic metals as Cd and Pb are absorbed to a small but significant extent. Presumably similar nonspecific mechanisms are involved in the absorption of these three polyvalent metals [122]. In
the form of organic complexes such as the glucose tolerance factor,
Cr(III) is absorbed relatively efficiently.

Table 2, based on the work of Donaldson and Barreras [123],
summarizes this discussion of Cr absorption with results obtained in
patients to whom labeled inorganic Cr was given orally or intraduo-
denally. Note especially (1) essentially complete excretion, that
is, close to zero absorption, of Cr(III), (2) small but significant
absorption of Cr(VI), and (3) the larger absorption of Cr(VI) after
duodenal than after oral administration, reflecting presumably gastric reduction of Cr(VI) to Cr(III). Tissue levels of Cr after exposure to Cr(VI) greatly exceed those following Cr(III) ingestion.
Spleen and kidney contained the highest amounts of Cr [124]. Not
surprisingly, this distribution can be greatly altered by the presence of various metal ligands. In plasma Cr circulates in part
bound to the iron-transporting protein siderophilin; the average
blood level of Cr in 20 normal subjects was reported as 2.9 μg/ml
[125]. The half-life of intravenously injected Cr(III) in plasma

TABLE 2

Urinary and Fecal Excretion of ^{51}Cr in Humans[a]

Compound administered	Percentage of radioactivity excreted			
	Oral administration		Duodenal administration	
	Feces	Urine	Feces	Urine
$^{51}Cr^{3+}Cl_3$	99.6 ± 1.8	0.5 ± 0.3	93.7 ± 4.5	0.6 ± 0.5
$Na_2^{51}Cr^{6+}O_4$	89.4 ± 2.6	2.1 ± 1.5	56.5 ± 11.7	10.6 ± 5.6

[a]Values are means ± SD.
Source: Based on results in Ref. 123.

is reported to equal approximately 6 hr [126]. Isotopic equilibration between plasma and tissue Cr can adequately be described in
terms of three tissue compartments, distinguished by their respective
turnover rates. Different organs contain various proportions of each
compartment.

Although some Cr may appear in feces following systemic administration, 80% or more of the total Cr excreted following its injection appears in urine. Because of the difficulty of distinguishing
between Cr(III) and Cr(VI) in biological samples, little is known
about the form in which Cr is excreted. In dogs about 10% of plasma
Cr circulates in diffusible form, complexed to small organic molecules. This Cr fraction is filtered at the glomerulus and is partially reabsorbed [127]. When very tightly bound in complexes such
as Cr-EDTA, the metal can no longer react with tubular receptors, so
that ^{51}Cr-EDTA can serve as a convenient marker for measurement of
the glomerular filtration rate [128].

8.3. Toxicity

Because they are only minimally absorbed across the intestinal barrier, and as they do not readily penetrate cells, inorganic compounds

of Cr(III) are generally not toxic or mutagenic. Following its
uptake into cells, Cr(VI) is rapidly reduced, and it is in the form
of Cr(III) that the metal reacts with DNA, proteins, and other cell
constituents. Chronic ingestion of Cr(VI) also produces relatively
few effects. Thus administration of K_2CrO_4 in drinking water (500
mg/liter) or of $ZnCrO_4$ (1% in food) to mature white mice did not
affect their general condition or reproductive capabilities [129].
Nevertheless, Cr(VI) compounds are generally corrosive, and skin
ulceration and sensitivity effects have been described [130]. In
large acute doses chromates or dichromates may be fatal. Damage is
caused especially to the proximal convoluted tubule in the kidney
[131].

The carcinogenicity of Cr compounds has been widely investigated in experimental animals and in humans. No carcinogenic
effects of ingested Cr have been reported. Injection of Cr compounds may lead to sarcomas at the site of administration [132].
The main evidence for carcinogenicity is based on surveys of workers
in the chromate industry who had been exposed to Cr-containing dusts
in the atmosphere. In contrast to the well-documented carcinogenic
effect of Cr in humans, attempts to induce bronchogenic carcinomas
by exposing animals to chromate dusts were not successful [133].

Effects on the respiratory system range from nasal perforation
[134] to changes in dynamic pulmonary function [135] and cancer [136].
An increased incidence of pulmonary carcinomas among workers in the
chromate industry has been reported from many countries, although
the precise extent and nature of the exposure in each case is not
always clear.

9. URANIUM

9.1. General Aspects

The toxicity of uranium compounds has been recognized ever since
large quantities of uranium ores began to be processed for the

extraction of radium. However, it was the study of nuclear fission
in World War II which led to the systematic analysis of the toxicological properties of the metal [137]. We will consider here the
chemical effects of uranium compounds on higher organisms; the radiation hazards associated with the unstable uranium isotopes will not
be further described. The high incidence of lung disease in uranium
miners results from radiation and silica exposure, rather than from
chemical toxicity of uranium.

Uranium (atomic number 92) is one of the actinide elements
and has the highest atomic weight of all generally recognized and
naturally occurring elements. It can exist in several valence
states (3, 4, 5, and 6), but only U(IV) and U(VI) are stable in
aqueous solution. The most important compounds of these two species
are oxides and halides; their toxicity depends on physicochemical
properties, on the rate and route of administration, on the presence
of toxic halides as in UF_6, and on other factors. Because of their
wide industrial use and relatively ready solubility, hexavalent U
salts of the uranyl ion $(UO_2)^{2+}$ have been most widely studied. The
uranyl ion readily forms complexes with organic and inorganic anions
in neutral solution. The complexes with bicarbonate and citrate are
especially stable and determine to some extent the fate of the uranyl
ion in vivo.

9.2. Absorption, Distribution, and Excretion

Absorption of U differs widely among various compounds, depending
on their physicochemical properties and the route of their administration. Both pulmonary and gastrointestinal exposures may be
important sources of the metal, but absorption from the lungs generally is more efficient than from the intestine; estimates of
pulmonary U absorption as high as 50% have been obtained in the
rabbit. Little is known of the mechanism of U absorption from the
intestine but, as with other heavy metals, fractional absorption is
quite low [138]. Various U compounds may also be absorbed through

the skin, and uranium is found in the bloodstream of rats following
dermal exposure to uranyl nitrate.

Distribution of U in the body depends on the nature of the
compound administered. After injection U is cleared fairly rapidly
from the blood, with uptake primarily into kidneys and bone. Uranyl
salts, unlike compounds of U(IV), are neither accumulated by the
liver to a significant extent nor recovered in bile or feces. The
main route of excretion of U(VI) is in urine.

9.3. Toxicity

Although effects of U compounds on a variety of isolated systems
can readily be demonstrated [139], the kidneys are the major target
organ in vivo [140]. The ready formation of bicarbonate complexes
of UO_2^{2+}, together with the additional observation that alkalosis
reduces the toxic effects [141], suggested that the bicarbonate
complex is filtered at the glomerulus and subsequently decomposed
upon acidification of tubular urine; UO_2^{2+} set free could then react
with renal tubule cells. The critical factors in this hypothesis
are urinary acidification and the relative absence of proteins to
complex U before it reaches the cell membranes.

While, as expected in this hypothesis, the proximal tubule
shows the major changes in U poisoning, a detailed analysis of
tubular function in rabbits following acute injection of uranyl
salts demonstrated functional lesions also in distal segments of
the tubules [142]. Injection of 0.2 mg U/kg as uranyl acetate
inhibits the active step in proximal secretion of p-aminohippurate
and the ability of the more distal segments to reabsorb NaCl [143];
effects of U on glomerular filtration rate and tubular anion transport are reversible. Rats and dogs are considerably more resistant
to U than are rabbits, but uranyl salts remain in general a useful
experimental tool for the production of acute renal failure [144].

As in the case of several other heavy metals, chronic low-
level exposure to U protects animals against a subsequent and

normally lethal dose of the metal. For instance, Haven [145] reported that intraperitoneal injection of 25 mg UCl_4/kg killed 9 out of 10 control rats, but only 1 out of 10 rats which had been pretreated with 11 daily injections of 1.65 mg/kg of the same compound. The mechanism of this acquired tolerance has not been elucidated and there is no evidence to suggest that it might involve induced synthesis of metallothionein, the low molecular weight metal-binding protein involved in the metabolism of cadmium, zinc, and other metals.

10. CONCLUSIONS

It is evident from the present review that individual metals in many respects possess unique toxicological characteristics. To some extent at least, this uniqueness may be more apparent than real. Each metal has been characterized in a somewhat different manner. Thus electrophysiological endpoints of toxicity have been intensively investigated only in the case of lead and, to a lesser extent, in the case of mercury and cadmium. Similarly, the peroxidative effects of only a few of the metals have been characterized. There is a need for more extensive comparative study of metal toxicity utilizing the same endpoints.

Another area in need of further study concerns the bioavailability of metals and the mechanisms whereby the body influences bioavailability. Evidence to date suggests that both lead and cadmium have the capacity to induce the formation of metal-binding proteins. The full significance of such potentially protective mechanisms is not fully understood.

Another area of potentially fruitful investigation concerns the effects of metals on the function of essential metals. Thus lead has been shown to materially influence the uptake and release of calcium by tissue preparations, an effect which has profound implications for normal physiological function, notably muscle contraction and the uptake and release of biogenic amines. Similar interactions with other essential metals are clearly in need of more intensive investigation. Finally, the effects of metals on the immune system [146] deserve greater attention than has been received in the past.

REFERENCES

1. R. A. Kehoe, *J. R. Inst. Public Health Hyg.*, *24*, 81 (1961); *24*, 129 (1961).

2. S. B. Gross, E. A. Pfitzer, D. W. Yeager, and R. A. Kehoe, *Toxicol. Appl. Pharmacol.*, *32*, 638 (1975).

3. B. J. Aungst and H. Fung, *Toxicol. Appl. Pharmacol.*, *61*, 39 (1981).

4. P. S. I. Barry and D. B. Mossman, *Br. J. Ind. Med.*, *27*, 339 (1970).

5. M. B. Rabinowitz, G. W. Wetherill, and J. D. Kopple, *J. Clin. Invest.*, *58*, 260 (1976).

6. E. J. O'Flaherty, P. B. Hammond, and S. I. Lerner, *Fund. Appl. Toxicol.*, *2*, 49 (1982).

7. P. B. Hammond, E. J. O'Flaherty, and P. S. Gartside, *Food Cosmet. Toxicol.*, *19*, 631 (1981).

8. M. R. Moore, P. A. Meredith, B. C. Campbell, A. Goldberg, and S. J. Pocock, *Lancet, 2*, 661 (1977).

9. S. Hernberg and J. Nikkanen, *Arch. Environ. Health, 21*, 140 (1970).

10. S. Piomelli, C. Seaman, D. Zullow, A. Curran, and B. Davidow, *Proc. Nat. Acad. Sci. U.S.A.*, *79*, 3335 (1982).

11. P. B. Hammond, R. L. Bornschein, and P. Succop, *Environ. Res.*, (in press).

12. G. W. Richter, Y. Kress, and C. C. Cornwall, *Am. J. Pathol.*, *53*, 189 (1968).

13. R. A. Goyer, *Curr. Top. Pathol.*, *55*, 147 (1971).

14. S. R. V. Raghavan, B. D. Culver, and H. C. Gonick, *J. Toxicol. Environ. Health, 7*, 561 (1981).

15. B. J. Stokvis, *Z. Klin. Med.*, *28*, 1 (1895).

16. M. R. Moore, in *Lead Toxicity* (R. L. Singhal and J. A. Thomas, eds.), Urban and Schwarzenberg, Baltimore, 1980, p. 79 ff.

17. A. Goldberg, P. A. Meredith, S. Miller, M. R. Moore, and G. G. Thompson, *Br. J. Pharmacol.*, *62*, 529 (1978).

18. M. D. Maines and A. Kappas, *Ann. Clin. Res.*, *8*, 39 (1976).

19. P. A. Meredith and M. R. Moore, *Biochem. Soc. Trans.*, *7*, 39 (1979).

20. S. Hernberg, M. Nurminen, and J. Hasan, *Environ. Res.*, *1*, 247 (1967).

21. S. Leikin and G. Eng, *Pediatrics, 31,* 996 (1963).

22. B. B. Gelman, I. A. Michaelson, and J. S. Bus, *Toxicol. Appl. Pharmacol.*, *45*, 119 (1978).

23. O. A. Levander, V. C. Morris, and R. J. Ferretti, *J. Nutr.*, *107*, 2135 (1977).

24. D. Holtzman, J. Shen Hsu, and M. Desautel, *Toxicol. Appl. Pharmacol.*, *58*, 48 (1981).

25. J. F. Rosen, R. W. Chesney, A. Hamstra, H. F. DeLuca, and K. R. Mahaffey, *N. Engl. J. Med.*, *302*, 1128 (1980).

26. D. R. Fraser, *Physiol. Rev.*, *60*, 551 (1980).

27. P. A. Meredith, B. C. Campbell, M. R. Moore, and A. Goldberg, *Eur. J. Clin. Pharmacol.*, *12*, 235 (1977).

28. A. P. Alvarez, S. Kapelner, S. Sassa, and A. Kappas, *Clin. Pharmacol. Ther.*, *17*, 179 (1973).

29. F. Buchthal and F. Behse, *Br. J. Ind. Med.*, *36*, 135 (1979).

30. A. M. Seppalainen, S. Tola, S. Hernberg, and B. Kock, *Arch. Environ. Health, 30,* 180 (1975).

31. K. Kostial and V. B. Vouk, *Br. J. Pharmacol.*, *12*, 219 (1957).

32. R. S. Manalis and G. P. Cooper, *Nature (London)*, *243*, 354 (1973).

33. G. P. Cooper and D. Steinberg, *Am. J. Physiol.*, *232*, C128 (1977).

34. D. Otto, V. Benignus, K. Muller, C. Barton, K. Seipl, J. Prah, and S. Schroeder, *Neurobehav. Toxicol. Teratol.*, *4*, 733 (1982).

35. D. A. Fox, J. P. Lewkowski, and G. P. Cooper, *Neurobehav. Toxicol.*, *1*, 101 (1979).

36. P. D. Hrdina, I. Hanin, and T. C. Dubas, in *Lead Toxicity* (R. L. Singhal and J. A. Thomas, eds.), Urban and Schwartzenberg, Baltimore, 1980, p. 273 ff.

37. R. L. Bornschein, D. Pearson, and L. Reiter, *CRC Crit. Rev. Toxicol.*, *8*, 43 (1980).

38. R. L. Bornschein, D. Pearson, and L. Reiter, *CRC Crit. Rev. Toxicol.*, *8*, 101 (1980).

39. L. Hastings, H. Zenick, P. Succop, T. J. Sun, and R. Sekeres, *Toxicol. Appl. Pharmacol.*, *73*, 416 (1984).

40. E. A. Creselius, *Environ. Health Perspec.*, *19*, 147 (1977).

41. J. P. Buchet, R. Lauwerys, and H. Roels, *Int. Arch. Occup. Environ. Health, 48,* 71 (1981).

42. J. P. Buchet, R. Lauwerys, and H. Roels, *Int. Arch. Occup. Environ. Health, 46,* 11 (1980).

43. R. Mappes, *Int. Arch. Occup. Environ. Health, 40,* 267 (1977).

44. T. Inamasu, A. Hisanaga, and N. Ishinishi, *Toxicol. Lett., 12,* 1 (1982).

45. G. Pershagen, B. Lind, and N.-E. Bjorklund, *Environ. Res., 29,* 425 (1982).

46. J. P. Buchet, R. Lauwerys, and H. Roels, *Int. Arch. Occup. Environ. Health, 48,* 111 (1981).

47. M. Vahter and J. Envall, *Environ. Res., 32,* 14 (1983).

48. J. B. Cavanagh, *Arch. Pathol. Lab. Med., 103,* 659 (1979).

49. P. B. Hammond, personal observation.

50. A. Heymans, J. B. Pfeiffer, R. W. Willett, and H. M. Taylor, *N. Engl. J. Med., 254,* 401 (1956).

51. C.-H. Tay and C.-S. Seah, *Med. J. Aust., 2,* 424 (1975).

52. J. S. Woods and B. A. Fowler, *Environ. Health Perspec., 19,* 209 (1977).

53. E. F. Baroch, in *Kirk-Othmer Encyclopedia of Chemical Technology,* Vol. 23, 3rd ed., John Wiley, New York, 1978, p. 673 ff.

54. E. G. Dimond, J. Caravaca, and A. Benchimol, *Am. J. Clin. Nutr., 12,* 49 (1963).

55. T. B. Wiegmann, H. D. Day, and R. V. Patak, *J. Toxicol. Environ. Health, 10,* 233 (1982).

56. A. R. Byrne and L. Kosta, *Sci. Total Environ., 10,* 17 (1978).

57. E. Sabbioni, E. Marafante, L. Amantini, L. Ubertalli, and C. Bivatarri, *Bioinorg. Chem., 8,* 503 (1978).

58. J. L. Johnson, H. J. Cohen, and K. V. Rajagopalan, *Biochem. Biophys. Res. Commun., 56,* 940 (1974).

59. H. Sakurai, S. Shimomura, K. Fukuzawa, and K. Ishizu, *Biochem. Biophys. Res. Commun., 96,* 293 (1980).

60. J. H. Tipton and J. J. Shafer, *Arch. Environ. Health, 8,* 58 (1964).

61. N. L. Kent and R. A. McCance, *Biochem. J., 35,* 837 (1941).

62. E. Sabbioni and E. Marafante, *Bioinorg. Chem., 9,* 389 (1978).

63. N. A. Talvitie and W. D. Wagner, *Arch. Ind. Hyg., 9,* 414 (1954).

64. A. W. Musk and J. G. Tees, *Med. J. Aust., 20,* 183 (1982).

65. G. L. Curran, D. L. Azarnoff, and R. E. Rolinger, *J. Clin. Invest., 38,* 1251 (1959).

66. J. Somerville and B. Davies, *Am. Heart J., 64,* 54 (1962).

67. L. C. Cantley, Jr., L. Josephson, and R. Warner, *J. Biol. Chem.*, *252*, 7421 (1977).

68. A. Schwartz, R. J. Adams, I. Grupp, G. Grupp, M. J. Holroyde, R. W. Millard, R. J. Solaro, and E. T. Wallick, *Basic Res. Cardiol.*, *75*, 444 (1980).

69. A. Huot, S. Muldoon, M. Pamnani, D. Clough, and F. J. Hady, *Fed. Proc.*, *38*, 1036 (1979).

70. T. Akera, K. Temma, and K. Takeda, *Fed. Proc.*, *42*, 2984 (1983).

71. P. M. Hudgins and G. W. Bond, *Pharmacology*, *23*, 156 (1981).

72. N. H. Stacey and C. D. Klaasen, *J. Toxicol. Environ. Health*, *7*, 139 (1981).

73. H. J. Drake, in *Kirk-Othmer Encyclopedia of Chemical Technology*, Vol. 15, 3rd ed., John Wiley, New York, 1978, p. 143 ff.

74. S. Jensen and A. Jernelow, *Nature (London)*, *223*, 753 (1969).

75. T. Rahola, T. Hattula, A. Cerolainen, and J. K. Miettinen, *Ann. Clin. Res.*, *5*, 214 (1973).

76. J. K. Miettinen, in *Mercury, Mercurials and Mercaptans* (M. W. Miller and T. W. Clarkson, eds.), Charles C. Thomas, Springfield, Illinois, 1973, p. 233 ff.

77. C. H. Walsh, *Environ. Res.*, *27*, 412 (1982).

78. I. R. Rowland, M. J. Davies, and J. G. Evans, *Arch. Environ. Health*, *35*, 155 (1980).

79. J. C. Gage, *Toxicol. Appl. Pharmacol.*, *32*, 225 (1975).

80. T. Norseth, and T. W. Clarkson, *Arch. Environ. Health*, *21*, 717 (1970).

81. L. W. Chang, *Environ. Res.*, *14*, 329 (1977).

82. L. W. Chang and H. A. Hartmann, *Acta Neuropathol.*, *20*, 316 (1972).

83. R. Von Burg, F. K. Northington, and A. Shamoo, *Toxicol. Appl. Pharmacol.*, *53*, 285 (1980).

84. N. H. Stacey and H. Kappus, *Toxicol. Appl. Pharmacol.*, *63*, 29 (1982).

85. M. Oka, D. F. Horrobin, M. S. Mankee, C. S. Cunnahane, A. I. Ally, and R. O. Morgan, *Toxicol. Appl. Pharmacol.*, *51*, 427 (1979).

86. F. A. X. Schanne, A. B. Kane, E. E. Young, and J. L. Farber, *Science*, *206*, 700 (1979).

87. L. Friberg, C. G. Elinder, T. Kjellström, and G. Nordberg, *Cadmium and Health*, CRC Press, Boca Raton, Florida (in press).

88. E. C. Foulkes (ed.), *Cadmium, Handbook of Experimental Pharmacology*, Springer-Verlag, Berlin (in press).

89. G. C. Cotzias, D. C. Borg, and B. Selleck, *Am. J. Physiol.*, *201*, 927 (1961).

90. I. Rabar and K. Kostial, *Arch. Toxicol.*, *47*, 63 (1981).

91. E. C. Foulkes, *Toxicology*, 1985, in press.

92. T. C. Harvey, J. S. McLellan, B. J. Thomas, and J. H. Fremlin, *Lancet*, *2*, 2969 (1975).

93. H. M. Perry and M. W. Erlanger, *J. Nutr.*, *112*, 1983 (1982).

94. T. Kjellström, C. G. Elinder, and L. Friberg, *Environ. Res.*, *33*, 284 (1984).

95. J. H. R. Kägi, and B. L. Vallee, *J. Biol. Chem.*, *236*, 2435 (1961).

96. M. Piscator, *Nord. Hyg. Tidskr.*, *45*, 76 (1964).

97. M. G. Cherian, R. A. Goyer, and L. S. Valberg, *J. Toxicol. Environ. Health*, *4*, 861 (1978).

98. E. C. Foulkes, *Am. J. Physiol.*, *227*, 1356 (1974).

99. E. C. Foulkes, *Toxicol. Appl. Pharmacol.*, *45*, 505 (1978).

100. E. C. Foulkes, *Trace Substances Environ. Health*, *19* (in press).

101. M. D. Enger, L. T. Ferzoco, R. A. Tobey, and C. E. Hildebrand, *J. Toxicol. Environ. Health*, *7*, 675 (1981).

102. A. Schnegg and M. Kirchgessner, *Z. Tierphysiol. Tierernaehr. Futtermittelkd.*, *36*, 63 (1975).

103. N. E. Dixon, C. Gazzola, R. L. Blakeley, and B. Zerner, *J. Am. Chem. Soc.*, *97*, 4131 (1975).

104. E. C. Foulkes and D. M. McMullen, *Toxicology*, 1985, in press.

105. E. Sabbioni and E. Marafante, *Environ. Physiol. Biochem.*, 132 (1975).

106. V. Eybl, J. Koutensky, J. Sykora, J. Svacinova, F. Merl, V. Senft, and I. Janousek, in *Ni. 3. Spurenelement Symposium* (M. Anke, H.-J. Schneider, and C. Bruckner, eds.), Friedrich-Schiller Universität, Jena, 1980, p. 315 ff.

107. F. W. Sunderman, Jr., *Ann. Clin. Lab. Sci.*, *7*, 377 (1977).

108. F. W. Sunderman, Jr., K. Kasprzak, E. Horak, P. Gitlitz, and C. Onkelinx, *Toxicol. Appl. Pharmacol.*, *38*, 177 (1976).

109. R. B. Ciccarelli, T. H. Hampton, and K. W. Jennette, *Cancer Lett.*, *12*, 34 (1981).

110. R. T. Haro, A. Furst, and H. L. Falk, *Proc. West. Pharmacol. Soc.*, *11*, 39 (1968).

111. P. D. Whanger, *Toxicol. Appl. Pharmacol.*, *25*, 323 (1973).

112. A. M. Ambrose, P. S. Larson, J. F. Borzelleca, and G. R. Hennigar, Jr., *J. Food Sci. Technol.*, *13*, 181 (1976).

113. P. H. Gitlitz, F. W. Sunderman, Jr., and P. J. Goldblatt, *Toxicol. Appl. Pharmacol.*, *34*, 430 (1975).

114. E. C. Foulkes and S. Blanck, *Toxicology*, *33*, 245 (1984).

115. K. S. Kasprzak, in *Ni. 3. Spurenelement Symposium* (M. Anke, H.-J. Schneider, and C. Bruckner, eds.), Friedrich-Schiller Universität, Jena, 1980, p. 345 ff.

116. North American Contact Dermatitis Group, *Arch. Dermatol.*, *108*, 537 (1972).

117. K. Kaaber, N. K. Veien, and J. C. Tjell, *Br. J. Dermatol.*, *98*, 197 (1978).

118. M. C. Herlant-Peers, H. F. Hildebrand, and G. Biserte, in *Ni. 3. Spurenelement Symposium* (M. Anke, H.-J. Schneider, and C. Bruckner, eds.), Friedrich-Schiller Universität, Jena, 1980, p. 87 ff.

119. W. Mertz, *Physiol. Rev.*, *49*, 163 (1969).

120. J. Alexander, J. Aaseth, and T. Norseth, *Toxicology*, *24*, 115 (1982).

121. F. L. Petrelli and S. DeFlora, *Appl. Environ. Microbiol.*, *33*, 805 (1977).

122. E. C. Foulkes, *Trace Substances Environ. Health*, *18*, 124 (1984).

123. R. M. Donaldson and R. F. Barreras, *J. Lab. Clin. Med.*, *68*, 484 (1966).

124. R. D. MacKenzie, R. U. Bjerrum, C. F. Decker, C. A. Hoppert, and F. L. Langham, *Arch. Ind. Health*, *18*, 232 (1958).

125. H. Nomiyama, M. Yotoriyama, and K. Nomiyama, *Am. Ind. Hyg. Assoc.*, *41*, 98 (1980).

126. H. F. Hopkins, *Am. J. Physiol.*, *209*, 731 (1965).

127. R. J. Collins, P. O. Fromm, and W. D. Collings, *Am. J. Physiol.*, *201*, 795 (1961).

128. B. D. Stacy and G. D. Thorburn, *Science*, *152*, 1076 (1966).

129. W. G. Gross and V. G. Heller, *J. Ind. Hyg. Toxicol.*, *28*, 52 (1946).

130. H. Royle, *Environ. Res.*, *10*, 141 (1975).

131. A. P. Evan and W. G. Dail, *Lab. Invest.*, *30*, 704 (1974).

132. W. W. Payne, *Arch. Ind. Health*, *21*, 530 (1960).

133. A. M. Baetjer, J. F. Lowney, H. Steffee, and V. Budacz, *Arch. Ind. Health*, *20*, 124 (1959).

134. T. F. Mancuso and W. C. Hueper, *Ind. Med. Surg.*, *20*, 358 (1957).

135. P. Bovet, M. Lob, and M. Grandjean, *Int. Arch. Occup. Environ. Health*, *40*, 25 (1977).

136. A. M. Baetjer, *Arch. Ind. Hyg. Occup. Med.*, *2*, 487 (1950).

137. C. Voegtlin and H. C. Hodge (eds.), *Pharmacology and Toxicology of Uranium Compounds*, McGraw-Hill, New York, 1949.

138. J. D. Harrison and J. W. Stather, *Radiat. Res.*, *88*, 47 (1981).

139. J. G. Mill, D. V. Vassallo, and V. P. Campanha, *Am. J. Physiol.*, *235*, H-295 (1978).

140. M. E. Bobey, L. P. Longley, R. Dickes, J. W. Price, and J. M. Hayman, Jr., *Am. J. Physiol.*, *139*, 155 (1943).

141. H. C. Hodge, *Arch. Ind. Health*, *14*, 43 (1956).

142. F. J. Bowman and E. C. Foulkes, *Toxicol. Appl. Pharmacol.*, *16*, 391 (1970).

143. K. Nomiyama and E. C. Foulkes, *Toxicol. Appl. Pharmacol.*, *13*, 89 (1968).

144. W. Flamenbaum, R. J. Hamburger, M. L. Huddleston, J. Kaufman, J. S. McNeil, J. H. Schwartz, and R. Nagle, *Kidney Int.*, *10*, S-115 (1976).

145. F. L. Haven, in *Pharmacology and Toxicology of Uranium Compounds* (C. Voegtlin and H. C. Hodge, eds.), McGraw-Hill, New York, 1949, p. 729 ff.

146. L. Treagan, *Met. Ions. Biol. Sys.*, *16*, 27 (1983).

7

Human Nutrition and Metal Ion Toxicity

M. R. Spivey Fox and Richard M. Jacobs
Division of Nutrition
Food and Drug Administration
Department of Health and Human Services
Washington, D.C. 20204

1. INTRODUCTION

The toxicity of 35 elements was reviewed to establish minimum tolerable levels of dietary minerals for domestic animals [1]. For most
of these elements there is extensive evidence that their toxicity
is markedly tempered by the dietary intake of other elements and a
diversity of other nutrients, such as protein and vitamins. Con-
versely, the toxic effects of some elements may depend to a large
extent on interference with the normal utilization and/or function
of essential minerals. The experimental data base for the National
Research Council (NRC) (see Ref. 1) is relevant to a similar evaluation for humans, although such a review has not been done.

There can be no argument that an adequately nourished person
is generally more healthy and less apt to be adversely affected by
a toxic exposure to a metal than a poorly nourished individual. The
desirability of prophylactic or therapeutic nutrient supplementation
requires evaluation for each metal and type of exposure.

Excess metal exposure may occur in industry, from environmental
contamination of water or foods, during therapy of a disease state,
from excessive intake of nutritional supplements, or from accidental
acute poisoning such as a small child's consumption of adult iron
tablets. This review will focus on the dietary exposure route, which
is probably the most sensitive to the nutritional status and dietary
intake of the subject. An important aspect of evaluating dietary
exposure to a toxic metal is the bioavailability of the element from
foods in which it is concentrated. The significance of nutrient-
metal interactions important to toxicity will be explored for the
essential elements selenium and zinc, a half-metal and a metal for
which extensive data exist.

2. NUTRIENT INTAKE AND NUTRITIONAL STATUS IN THE UNITED STATES

Data on large numbers of subjects have recently become available for
these two subject areas. The studies include the Nationwide Food

Consumption Survey (NFCS) [2] of the U.S. Department of Agriculture
and the Second National Health and Nutrition Examination Survey
(NHANES II) of the U.S. Department of Health and Human Services.
For several years the Food and Drug Administration has conducted the
Selected Minerals in Food Survey (SMIFS) based on the food consumption of the young adult male [3]. This survey has been expanded to
include eight age-sex groups ranging from infants to older adults.

The NFCS data are based on dietary intake records from 36,100
subjects. They show average daily intakes well below the Recommended
Dietary Allowance (RDA) [4] for calcium, magnesium, and iron for some
groups, particularly women. The SMIFS diets were originally developed
from 1965 food consumption data and were recently revised based on
1976-1980 data. These diets represent typical intakes for each age
group. Deficient intakes are indicated for zinc, iron, and copper,
primarily in the younger age groups. Detailed evaluation of NHANES II
data on mean corpuscular volume, transferrin saturation, erythrocyte
protoporphyrin, serum transferrin, and hemoglobin indicated iron deficiency less extensive than one might expect from the NFCS data [5].
Data were from 18,981 subjects, except that transferrin was determined
in a smaller number of subjects.

Markedly inadequate intakes of vitamin B_6 by most age groups
were also shown by the NFCS data. The Life Sciences Research Office
(LSRO) evaluated the NHANES II data for serum and red blood cell (RBC)
folate levels in 2910 and 2425 individuals, respectively [6]. The
serum values indicated low status for 1/10 to 1/4 of adult males and
approximately 1/10 of females beginning at 10 years of age. Red blood
cell values indicated smaller segments of the population with low
folate status. An LSRO evaluation of serum zinc values for 14,770
persons 3-74 years of age from the NHANES II data identified many
factors that may affect serum zinc levels independent of zinc status
[7]. After taking into consideration factors that could have caused
spurious values, they concluded that less than 2% of the males and
3% of the females had low serum zinc. They also concluded that serum
zinc was not a satisfactory measure of zinc status.

Intakes of nutrients in excess of requirements can affect the utilization of other essential nutrients, as well as that of toxic levels of metals. The NFCS data showed excessive intakes of protein for all age groups, as much as 150-200% of the RDA. There were smaller excessive intakes (about 120-150% RDA) of phosphorus, vitamin A, riboflavin, vitamin B_{12}, and vitamin C. In most cases the elevated intakes were greater for males than for females. The SMIFS data indicated moderate excess intakes of sodium and potassium but markedly high levels of iodine ranging from 213 to 1152% of the RDA. There was an overall trend to lower levels by 1982. The highest source of iodine was dairy products, as a consequence of iodine compounds used in feed supplements, veterinary medications, and cleaning solutions. Grain and cereal products were also high in iodine.

Consumption of vitamin-mineral supplements in the United States is widespread. About one-fourth to one-half of the NFCS subjects regularly used supplements. The highest use was by children under 5 years of age. In a separate telephone interview survey of 2991 adults, 43% of those surveyed consumed 1-14 products daily [8]. Vitamin C was the supplement most widely used. High-dosage consumption was greatest for the water-soluble vitamins. Studies similar to the above have been conducted in other countries, but space precludes summary in this review.

3. SELENIUM

The nutrition, metabolism, and toxicity of selenium are integrally related to many other nutrients and nonnutrient substances. This review will be restricted to those aspects of selenium nutrition and toxicity with obvious pertinence to humans. Other aspects of selenium metabolism, such as interactions between selenium and nonnutrient substances, have been reviewed [9-12].

3.1. Essentiality, Function, Effects of
      Deficiency, and Requirements

Selenium is an essential nutrient for most, if not all, animals
[13,14] and for some microorganisms [15]. The only well-characterized
function of selenium in animals is as part of the enzyme glutathione
peroxidase (GSH-Px) [16,17]. Selenium, as selenocysteinyl residues,
is present in stoichiometric amounts in GSH-Px, 1 g-atom of selenium
for each of four identical subunits [18-20]. The GSH-Px in the cyto-
sol of the cell and the mitochondrial matrix space reduces hydroper-
oxides, primarily H_2O_2. These functions in the hydrophilic regions
are equivalent to those of vitamin E in the lipophilic regions.
Other non-selenium-dependent glutathione peroxidases also exist [21].
Other selenoproteins have been observed in animal tissue; however,
no biochemical function has been ascribed [22-24].

     The edible portions of most plants contain selenium; however,
the concentrations vary greatly, depending on the amounts of selenium
in the soil. In several parts of the world nutritionally inadequate
amounts of selenium in forage and feed resulted in selenium deficiency
in livestock [14], which required supplemental selenium to alleviate
the problems [24,25].

     Schwarz and Foltz [13] demonstrated that selenium could prevent
dietary liver necrosis in rats fed a vitamin E-free diet. Selenium
deficiency in the presence of inadequate vitamin E causes several
disease states in livestock: exudative diathesis and pancreatic
fibrosis in poultry, muscular dystrophy in lambs and calves, and
hepatosis dietetica in pigs [14]. Nearly all signs of selenium deficiency were shown to be either partially or completely ameliorated
by vitamin E [26]. The discovery by Rotruck et al. [16] that selenium
functioned in GSH-Px revealed a role for selenium distinct from that
of vitamin E. Selenium is necessary for the growth and fertility of
most animal species.

     The selenium content of human foods and diets varies greatly.
The U.S. population appears to have adequate intakes of selenium

[3,27,28], whereas areas of Finland, New Zealand, and China have experienced endemic selenium undernutrition [29-31] and deficiency [32-35]. Areas of China are the only places where intakes were so low as to elicit severe signs of selenium deficiency in humans, which caused cardiomyopathy and death [35]. Individuals maintained on chemically defined diets [36] or total parenteral nutrition have also exhibited severe signs of selenium deficiency [37].

A human requirement for selenium has not been established; however, the NRC has established estimated safe and adequate daily intakes of selenium, 50-200 µg/day for adults [4]. Balance studies indicated that the human requirement approximates 70 µg/day [38]. Typical intakes for the United States ranged from 60 to 150 µg/day [3,27,28], whereas intakes in New Zealand and Finland were less than 60 µg/day. In areas of China where endemic selenium deficiency (Keshan disease) occurs, intakes averaged 11 µg/day [33,34].

3.2. Absorption, Metabolism, and Excretion

Selenium metabolism has been reviewed in depth elsewhere [14,26,39]. This discussion is limited to the forms of selenium most prevalent in foods at typical levels and the forms in dietary supplements.

The most important forms of selenium in foods are selenomethionine and selenocysteine (selenocystine). The forms in supplements include selenite, selenate, and yeast-borne selenium, which is purported to be selenomethionine. The forms of selenium actually present in supplements may differ from those listed on the label. Products claiming to be selenite may contain elemental or other forms of selenium, owing to chemical reactions with other ingredients in the supplement. Coadministered supplements containing ascorbic acid may also react with selenite before intestinal absorption. Most products containing yeast-borne selenium have not been rigorously characterized. Sometimes they contain elemental selenium or volatile forms of selenium.

The absorption of selenium compounds is highly efficient over
a wide range of intakes. Absorption is largely dependent on the form
of selenium and the presence of other interacting substances in the
diet [26]. The absorption of selenomethionine mimics that of methionine [40], and the absorption of selenite and selenate mimics that
of sulfite and sulfate, respectively. In rats selenomethionine and
selenite were absorbed to a similar extent in one study [41], whereas
others [26] demonstrated variable absorption. With low to moderate
intakes of selenium [14], tissue levels are largely controlled by
urinary excretion of selenium.

In the postabsorptive metabolism of selenium compounds selenite
is a common intermediate in the metabolism of selenate and organic
forms of selenium, such as selenomethionine and selenocysteine [42].
Selenate is reduced first to selenite and then to selenide, an important transport form of selenium. Selenides may be either methylated
or react directly with proteins and other thiols [39].

The selenium in selenomethionine may be converted to selenite
[42] or may replace methionine in tissue proteins [40,43], particularly when methionine is limiting in the diet. With excess dietary
methionine a greater proportion of selenomethionine is rapidly converted to selenocysteine and then selenite [44], rather than incorporated into tissue protein.

Incorporation of selenium in the selenocysteinyl residues of
GSH-Px apparently arises from posttranslational reactions of selenide
with the apoenzyme [45]. The biosynthesis of selenomethionine and
selenocysteine from selenite apparently does not take place in non-
ruminants [46]. At low to moderate intakes, selenium as selenite,
selenate, selenomethionine, and selenocysteine is at least partially
converted to trimethylselenonium ion by transmethylation of selenide
[40]. This is a highly soluble, relatively nontoxic form of selenium
which is excreted in the urine. Other, as yet uncharacterized, forms
of selenium in the urine are also important excretory metabolites at
low levels of selenium intake. Trimethylselenonium ion as an excretory product is probably more important at toxic intakes of selenium

than at low intakes [47]. At moderate to toxic intakes of selenium, dimethylselenide also becomes an important excretory metabolite of selenium. This highly volatile form of selenium is excreted via expired air and is responsible for the garliclike odor usually associated with either acute or chronic selenosis. Dimethylselenide is thought to be responsible for the pulmonary signs associated with chronic selenosis [48].

3.3. Selenium Toxicity in Animals

Ingestion of selenoamino acids in plants has led to acute and chronic selenium [48]. This highly volatile form of selenium is excreted via expired air and is responsible for the garliclike odor usually associated with either acute or chronic selenosis. Dimethylselenide is thought to be responsible for the pulmonary signs associated with chronic selenosis.
be acutely or chronically affected. In the initial stages of the disease animals wander and have uncertain gait, impaired vision, and loss of appetite. In later stages animals suffer from weakness in the front legs, respiratory failure, paralysis, and death. Liver necrosis, nephritis, hyperemia, and ulcerations of the upper intestinal tract are characteristic lesions in these animals [12,14].

Alkali disease is a chronic disorder arising from consumption of seleniferous grains containing lower levels of selenium, 5-40 ppm, primarily in the form of protein-type selenoamino acids such as selenomethionine and selenocysteine. Toxicities in experimental animals fed selenite or selenate are more characteristic of alkali disease than blind staggers [12]. Alkali disease in horses, swine, and cattle is characterized by rough hair coat, lack of vitality, loss of appetite, deformation of hoofs, loss of hoofs and hair, lameness, and emaciation. Liver cirrhosis, atrophy of the heart, hemolytic anemia, and death may occur in more advanced cases.

Rabbits fed 30 ppm selenium in the diet for 50 days showed no ill effects, but they then rapidly deteriorated with enteritis and

liver and kidney damage [51]. Hamsters that received 12 ppm selenium
in their drinking water lost weight [52]. The minimum lethal dose
for injected selenite ranged from 3.25 to 5.7 mg of selenium per
kilogram body weight [12]. Values for selenate, selenomethionine,
and selenocysteine fell within the same range.

Several workers [6,14,49,53-55] have studied the toxic effects
of selenium in the rat. At subacute exposures rats exhibited reduced
food intake, reduced growth, a marked progressive anemia, and pronounced pathological changes in the liver. They developed a characteristic hunched posture and occasionally had paralysis of the hind
quarters. Reproductive organs failed to develop fully and intestinal
hemorrhages were common.

Similar signs were observed in chronically exposed rats. Animals developed a progressive anemia and died when the hemoglobin
level reached approximately 2.0 g/dl blood. Characteristic changes
occurred in the liver, such as atrophy, necrosis, cirrhosis, and
hemorrhage. The liver often appeared to be the source of fatal hemor-
rhaging. Levels of selenium in drinking water that produced chronic
toxicity were approximately 2-3 ppm for selenite or selenate; 6-9 ppm
resulted in mortality [56]. Selenium at elevated levels decreased
fertility in most species. Rats and other species exhibited teratogenic effects [57-60].

The toxicity of selenium varies with the chemical form. Hydrogen selenide is by far the most toxic form; 1-4 µg/liter inhaled air
produces pulmonary edema and death in rabbits after 8-12 hr of exposure [54]. Apart from industrial exposures, hydrogen selenide poisoning rarely occurs. The extent to which endogenous hydrogen selenide is involved in selenite or selenate toxicity is not known. Some
selenium-containing supplements intended for human use contain volatile, noxious forms of selenium that may be hydrogen selenide [61,62].

Among the forms of selenium with practical importance to humans,
selenite has been the most extensively studied in animals. Less information exists for selenomethionine and selenocysteine. The toxicities of selenite and selenate are similar. Selenite and selenomethionine have similar toxicities when administered over long

periods. The onset of selenomethionine toxicity can be modified by
methionine intake and status. At low methionine intakes tissue
selenium levels in animals fed selenomethionine increased manyfold
over their selenite-fed controls, which substantially delayed the
toxic effects [63]. Methionine reduced the toxicity of selenite,
perhaps by contributing to a methyl donor pool depleted by reductive
methylation of selenite [39]; however, selenite is known to inactivate
methionine adenosyltransferase [64]. The protective effects of methionine are largely dependent on the presence of vitamin E [65].

High-protein diets also protect against selenite toxicity [66],
whereas low-protein diets tend to exacerbate selenium toxicity.
Various sources of protein elicit various protective effects, which
may be attributable to the methionine and sulfur content of the protein [66]. Dietary sulfate was partially protective against the
toxic effects of selenite and selenate, but less effective against
those of organic forms of selenium [67]. Linseed oil meal protected
against selenite toxicity [68]; the effects were attributed to two
cyanogenic glycosides rather than to protein [69].

3.4. Selenium Toxicity in Humans

Selenium toxicity from nondietary sources has been reviewed [48,70].
Chronic industrial exposures caused pulmonary signs (garliclike
breath, bronchitis), intestinal discomfort, nausea, lassitude, and
irritability. Teratogenic effects are possible at levels encountered
in industrial exposures to selenium [71,72].

Severe selenium toxicity, leading to death, was reported in two
individuals [73,74]. A 71-year-old man suffered a massive, fatal
heart attack after years of exposure to elemental selenium in the
workplace. Very high levels of selenium were found in many tissues,
such as the lungs, hair, and nails. Many tissues contained apparent
deposits of elemental selenium [73]. The second case, an acute
exposure resulting in death, was that of a 3-year-old boy who had
consumed an unknown quantity of a gun-blueing compound that contained

1.8% selenious acid [74]. Garliclike breath was detected and tissue
analysis confirmed the cause of death as selenium toxicity.

A teenage girl who ingested 22.3 mg of selenium/kg body weight
as sodium selenate exhibited abnormal electrocardiogram waves 3 days
later [75]. She had abnormal liver function for about 1 week but
recovered without apparent ill effects.

Human exposures in excess of the estimated safe and adequate
daily dietary intake range established by the NRC have been reported
without toxicity. Sakurai and Tsuchiya [76] reported that Japanese
fishermen consume 500 µg of selenium per day or more, largely from
seafood, without ill effects. Schrauzer and White [77] reported
that two individuals consuming 350 and 600 µg of supplemental selenium per day in addition to dietary sources for 18 months were devoid
of adverse clinical or laboratory signs. A selenium-yeast preparation
was the source of the selenium.

Venezuelan children living in high-selenium areas had elevated
blood selenium levels [78,79]. Although some children exhibited
signs of mild selenosis (nausea, loose hair, and nail changes), these
changes could not be attributed directly to selenium. Many of the
Chinese subjects studied by Yang et al. [50] had estimated intakes
of 750 µg/day without apparent toxicity.

In areas of the United States where selenosis was prevalent in
animals, no specific signs of selenium intoxication were noted in
human residents [80], although 31% complained of gastrointestinal
disturbance that may have been attributable to selenium exposure.
Lemly and Merryman [81] associated dermatitis, dizziness, lassitude,
and depression with chronic selenium toxicity in people from this
same area. These symptoms were reported to be relieved by consumption of a low-selenium diet.

Ironically, China has been the site of both endemic chronic
selenium toxicity and severe selenium deficiency in humans. Yang
and co-workers [50] reported on an outbreak of selenosis caused by
drought and crop failure which forced villagers to consume increased
amounts of vegetables and corn containing high levels of selenium.

The selenium in the plant foods apparently arose from burning high-selenium coal and using lime fertilizer. During years when selenosis was most prevalent, approximately 50% of the inhabitants of affected villages were intoxicated with selenium. After the peak years of selenosis the average daily dietary intake was estimated to be 5 mg. Signs and symptoms of selenosis included loss of hair and nails, skin rash, gastrointestinal disturbances, and effects on the central nervous system and possibly teeth. Selenium levels in hair and urine were approximately 100-fold higher in these individuals than in those living in selenium-adequate areas. Blood selenium content was roughly 30-fold higher in the area with chronic selenosis. Although no liver damage was reported, severe effects on the nervous system were observed in 18 people from one village. One person developed hemiplegia and later died. The form of selenium in the diets was not identified; however, some of the signs resembled those of blind staggers, so a fraction of the selenium may have been in the form of non-protein-type selenoamino acids. Although no food intake data were presented, the authors indicated that the severity of the selenosis may have been related to the consumption of diets containing low levels of protein.

Yang et al. [50] also reported that a 72-year-old man suffered mild signs of selenium toxicity after consuming 2 mg of sodium selenite per day (presumably 0.9 mg of selenium) for more than 2 years. Tissue levels of selenium were far lower than those in individuals living in seleniferous areas without selenosis.

Recently, selenium intoxication resulted from the ingestion of a dietary selenium supplement containing approximately 27 mg of selenium per tablet, 182 times the amount claimed on the label [82]. Some signs of selenium toxicity were noted in all 13 of the identified users of this product. Signs and symptoms included nausea, vomiting, garlic breath, dermatitis, and loss of hair and nails. A 57-year-old female consumed a total of 2387 mg of selenium over a 77-day period with loss of nails and hair, but without apparent liver damage. Visual examination of these tablets revealed pinkish red flecks, which probably were elemental selenium. Although much of the selenium present in the tablet remained water soluble and was therefore probably bioavailable, it was remarkable that no liver damage was apparent. Consumption of these amounts of selenium can be fatal [48]. The lack of more serious effects was attributed to the simultaneous consumption of high-potency vitamin and mineral supplements containing vitamins A, C, and E [83,84].

Although there is no compelling evidence that human selenium toxicity is substantially altered by nutrient intake or status, it can be inferred that this is likely. Many of the biochemical pathways for selenium metabolism in humans are similar to those in animals, where protective effects of nutrients have been observed. Similar metabolic effects have been demonstrated in human nutritional studies utilizing selenate, selenite, and selenomethionine [85,86].

3.5. Significance of Selenium-Nutrient Interactions for Humans

Currently the potential risk for endemic dietary selenium toxicity from environmental circumstances is small. This is in part due to the identification of soils and foods with a high selenium content and food distribution practices (at least in the United States) that tend to minimize the extremes of nutrient intake. However, changing agricultural practices and environmental circumstances can cause the mobilization of soil selenium, resulting in dramatic increases of selenium in forage and food [48].

The potential risk of selenium toxicity arising from the use of dietary supplements is somewhat higher. Although in the United States maximum potencies of selenium-containing dietary supplements are limited to 200 µg/day [87], errors in formulation or consumption beyond the recommended amount can lead to excessive intakes and toxicity.

According to nutrient intake data, the levels of those nutrients known to modify selenium toxicity beneficially (protein, methionine, and vitamin E) exceeded the requirement. Although no intake data

were available specifically for methionine, the level of methionine
intakes probably exceeded the requirement because protein intake was
high [2]. Vitamin E intake is thought to be adequate in the United
States [4]. Moreover, 34% of adult supplement users in a U.S. survey
consumed vitamin E supplements [8]. One might conjecture that a
vegetarian-type diet, devoid of rich sources of methionine such as
eggs and milk, might provide the least protection against excessive
selenium intakes.

In regard to the potential toxicity from various forms of
supplemental selenium, one might foresee physiological conditions
that could increase the toxicity of selenomethionine compared to
that of inorganic forms of selenium or selenocysteine. Increased
protein requirement (such as in infants, children, and pregnant or
lactating women) or low protein (methionine) intake increases the
incorporation of selenomethionine into body proteins, leading to
much larger body stores of selenium. One would expect that these
individuals would be more at risk to selenium toxicity from endogenous stores of selenium during periods of rapid protein catabolism
caused by debilitating disease or rapid weight loss.

There is an incomplete understanding of the metabolism of
selenium in various forms at both deficient and excessive intakes
and of interacting dietary factors. The requirement and maximum
safe level(s) of selenium are not known precisely. Assessment tests
for human selenium status at low and high intakes are needed; however, the GSH-Px level in platelets may ultimately provide some
measure of selenium status at suboptimal intakes of selenium [85,88].
These problems need to be investigated further in both humans and
animals.

4. ZINC

4.1. Essentiality, Function, Effects of
Deficiency, and Requirements

Zinc is essential for all microorganisms, plants, and animals that
have been studied. Zinc has been reported to function in 52 enzymes

(202 from all sources), which include 5 oxidoreductases, 9 trans-
ferases, 30 hydrolases, 5 lyases, 1 isomerase, and 2 ligases [89].
The functions of zinc involve catalytic, structural, and regulatory
processes. The zinc enzymes affect the metabolism of carbohydrate,
lipid, protein, and nucleic acid (both RNA and DNA). It appears
that zinc has a role in gene regulation, possibly through zinc-
dependent enzymes [90]. There is evidence that zinc also plays a
role in the maintenance of membrane structure and function [91].

The effects of zinc deficiency have many similarities in the
numerous species that have been studied. Defects that have been
observed include growth depression, skin lesions, esophageal lesions,
abnormal morphology of hair and feathers (ultimate loss), skeletal
abnormalities, hypogonadism (studied predominantly in males) and
delayed maturation, impaired/aberrant senses of taste and smell,
abnormal dark adaptation, impaired brain function, impaired immune
function, atrophy of the thymus and lymphoid tissue, lymphopenia,
delivery problems at birth, and a variety of defects in the newborn.
These effects have been reported in an extensive literature, much of
which has been summarized in monographs and reviews [14,92,93].

The RDA for adult men and women is 15 mg/day, with an additional
5 and 10 mg during pregnancy and lactation, respectively [4]. For
infants the RDA is 3 mg/day from birth to 6 months of age, 5 mg/day
to 1 year of age, and 10 mg/day for children 1-10 years of age. The
bioavailability of dietary zinc can be affected by other dietary
components, such as fiber, phytate (particularly with high calcium),
and other minerals [94,95].

4.2. Effects of Excess Zinc in Animals

In early studies of zinc toxicity, adverse effects usually were not
observed at dietary levels below 1000 ppm. This led to a long-held
belief that zinc was relatively nontoxic. Since stock diets were
used in those studies, the fiber, phytate, and probable high mineral
content undoubtedly contributed to the limited adverse effects of
zinc.

Growth reduction usually accompanies high zinc intake. Grant-Frost and Underwood [96] observed that decreased food consumption and utilization contributed to the slowed growth rate. They thought that the unpalatability of the diet and refusal of animals to eat were important factors contributing to the adverse effects of zinc. Marginal copper intake increased the adverse effects of zinc [97,98]. Achromatrichia was observed in rats fed only the requirement levels of copper with 1000 ppm zinc [99]. Decreased feather pigmentation in young Japanese quail occurred with 31 ppm supplemental zinc at 1 ppm dietary copper (requirement 1.5 ppm) and 2000 ppm with 3.6 ppm copper [98].

A microcytic, hypochromic anemia was observed frequently in early studies of animals fed diets with 1000 ppm or higher zinc [96,100]. Increased erythrocyte fragility was also found [101]. The anemia was at least partially ameliorated by supplements of copper [102,103] or iron [102,104]. Cox and Harris [104] prevented the anemia with copper alone, whereas combined supplements of copper and iron were required to prevent the anemia in other studies [102, 103]. With copper-deficient diets decreased hemoglobin occurred with 400 ppm supplemental zinc or less in chicks and mice, but not in rats [97,105-107]. Supplemental calcium prevented the anemia caused by a diet with a high zinc-to-copper ratio in mice [108]. In rats fed 6300 ppm zinc, adverse effects on hemoglobin were exacerbated by low dietary calcium [109].

Bone abnormalities in rats fed 2000-10,000 ppm zinc included decreased weight and ash and abnormal morphology [110,111]. Stewart and Magee [111] prevented the defects with supplemental calcium and phosphorus. Enlargement of epiphyseal regions of the long bones was the first sign of zinc toxicity in foals [112]. Supplemental ascorbic acid, which is known to decrease copper absorption, and excess zinc each caused perosis in young Japanese quail fed a diet marginal in copper [113]. The combined supplements had greater effects than either alone; adverse effects occurred with as little as 160 ppm zinc plus 2.5 g of ascorbic acid per kilogram diet.

In most of the above studies of excess zinc intake, data were
also presented that showed elevated levels of zinc in many tissues.
The elevated liver zinc was primarily in the soluble fraction, bound
to metallothionein [114]. Sequestration of zinc in metallothionein
probably serves a temporary storage and detoxification function.
There have also been reports of marked changes in concentrations of
zinc metalloenzymes in many tissues following high zinc intake.

Changes in tissue mineral concentrations have been reported,
including decreased ash in bone [111] and decreased iron in liver
[98,104]. Decreased manganese was found in the liver of Japanese
quail fed between 250 and 2000 ppm zinc [98]. There have been reports of decreased concentrations of copper in the liver [98,104].
Decreased ceruloplasmin has been frequently observed, and Campbell
and Mills [99] found decreased ceruloplasmin with only 300 ppm dietary
zinc when rats received a diet containing the required level of copper
only. Changes in other copper and iron enzymes have been reported
with excess zinc intake.

Large amounts of zinc have been shown to decrease copper absorption [115]. In rats fed a diet containing 900 ppm zinc, copper-64
absorption was decreased; however, distribution of copper among tissues within the carcass was unaffected [116]. There were high concentrations of copper-64 in the intestinal mucosa, most of which was
bound to metallothionein. It is believed that the decreased copper
absorption is due to induction of metallothionein by zinc followed
by displacement of zinc on metallothionein by copper, which has a
greater affinity for metallothionein than zinc [116]. This trapped
copper appears to be essentially unavailable for transport from the
mucosa.

The principal metabolic effect of excess zinc is related to
lipids. On the basis of data in rats, Klevay [117] proposed that
consumption of diets with high ratios of zinc to copper resulted in
hypercholesterolemia. Copper deficiency ultimately was shown to be
the primary condition producing hypercholesterolemia [118]. The
effect of copper on cholesterol metabolism has been further confirmed

in responses to several food components and one drug that produce
negative and positive effects on copper status [119].

4.3. Effects of Excess Zinc in Humans

The undesirable taste of zinc in water is detectable at 15 ppm and
is very marked at 40 ppm. At higher levels zinc is an emetic and
can also cause diarrhea. There have been occasional reports of
acute zinc toxicity from foods cooked or held in galvanized containers. Accompanying cadmium in the vessels may have contributed
to these illnesses. Other acute episodes of zinc poisoning have
occurred in the workplace, primarily from inhalation of zinc fumes
or dust.

The effect of excess zinc (140 mg/day as the sulfate) on
calcium absorption was determined in adult males following a dose
of calcium-47 [120]. During low calcium intake (230 mg/day), supplemental zinc caused decreased calcium-47 levels in the plasma and
increased fecal excretion. There were no effects of zinc when calcium intake was normal (800 mg/day).

The effects of elevated zinc intakes on copper balance have
been studied by several workers. Greger et al. [121] found that
increasing the zinc intake of adolescent women from 11.3-11.6 to
14.5-14.8 mg/day caused increased fecal copper loss; however, copper
retention was unchanged. In contrast, a 27% decrease in apparent
copper absorption occurred in elderly subjects when dietary zinc
was increased from 7.8 to 23.3 mg/day [122]. Sandstead [123] reported that copper requirements for balance increased from 0.89 to
1.64 mg/day when zinc intake was increased from 5 to 20 mg/day in a
diet containing 80 g/day protein.

The above effects of zinc on copper status, which occur within
"normal" intake ranges of both elements, are in marked contrast to
effects on copper observed with much higher exposure to zinc in clinical settings. The most severe effects of zinc toxicity occurred in
dialysis patients when the tap water used in dialysis had contacted

galvanized iron, which usually had been recently installed [124, 125]. Nausea, vomiting, and fever occurred in some instances and all patients became severely anemic. Ten patients with 7.05 g/dl hemoglobin had further declines to 5.07 g/dl over a 2-month period (p < 0.01) despite transfusions. A teenage male exposed to zinc during home dialysis required blood transfusions, which had transient benefits. His hemoglobin was about 3 g/dl and RBC survival time was 3 days. Elimination of the high amounts of zinc in the water solved the toxicity problems.

A premature infant who developed acrodermatitis enteropathica (a genetic defect resulting in decreased zinc absorption) was in clinical remission at 17 weeks following 30 mg/day zinc for 3 weeks [126]. She had low serum copper, neutropenia, and radiological evidence of copper deficiency. Although urinary zinc was very high, the dose was excessive, probably because of her small size and continuation of breast feeding, which permits higher zinc absorption than formula feeding in this disease.

Porter et al. [127] produced marked improvements with supplemental zinc in patients with nonresponsive celiac disease. One of six patients, who continued to take 660 mg of zinc sulfate (probably 150 mg of zinc) per day after discharge, developed severe anemia after 2 months. She had neutropenia, low hemoglobin, and low serum iron and copper. A transfusion of erythrocytes and 4 mg of oral copper sulfate per day returned the blood parameters to normal after 4 weeks.

Microcytosis and leukopenia were observed in a sickle cell anemia patient following administration of 150 mg/day zinc as the sulfate for 2 years [128]. All adverse changes were reversed by supplemental copper. Of 13 other patients who had received zinc for 4-24 weeks, 7 had markedly reduced ceruloplasmin levels but no other changes. All values had been normal on admission. Abdulla [129] administered 45 mg/day zinc to seven healthy adults for 12 weeks. There were small increases in plasma zinc and decreases in plasma copper.

Freeland-Graves et al. [130] adjusted zinc-to-copper ratios
for 32 healthy women to 3, 13, 21, and 64 by zinc supplementations
of 0, 15, 50, and 100 mg/day zinc as the acetate, respectively.
After 60 days there were no uniform or sustained effects of the
supplements on plasma cholesterol or high-density lipoprotein
cholesterol. Plasma cholesterol was significantly correlated with
dietary copper ($r = -0.21$, $p < 0.009$). Contrasting results were
reported by Hooper et al. [131] in a 5-week study of 20 normal men.
Twelve received 80 mg/day zinc as the sulfate. After 5 weeks of
zinc supplementation high-density lipoprotein cholesterol levels
declined significantly. This change persisted to 7 weeks, but by
16 weeks the value returned to baseline. There were no changes in
the placebo control subjects. Total serum cholesterol, triglyceride,
and low-density lipoprotein cholesterol in the zinc-supplemented
subjects did not change.

Experimental copper depletion was studied in a healthy 29-year-
old man who consumed a diet of conventional foods that supplied 0.83
mg/day copper [132]. The diet was otherwise adequate, except for
calcium and zinc, which were supplied as gluconate salts to provide
total intakes of 840 and 13.7 mg of calcium and zinc, respectively.
Cholesterol intake was 200 mg/day and the ratio of linoleate to
saturated fatty acids was 0.5. The study began with a 43-day control phase during which the subject received the diet supplemented
with 0.5 mg/day copper. During the 105-day depletion phase there
were losses of copper in urine and feces that exceeded intake.
There were significant declines in plasma copper and ceruloplasmin
and in superoxide dismutase activity of the erythrocytes. There
was an increase in plasma cholesterol and a smaller increase in low-
density lipoprotein cholesterol; there were no changes in plasma
triglyceride, high-density and very-low-density lipoprotein cholesterol, and hematological indices. After 29 days of repletion with
4 mg/day copper, given as an aqueous solution of copper sulfate, all
parameters were in the normal range.

4.4. Significance of Zinc-Nutrient Interactions for Humans

This brief review of the effects of excess zinc in humans provides data that confirm the results of zinc-nutrient interactions observed in experimental animals. The most sensitive relationship and the one that appears to be of greatest practical importance to humans is the zinc antagonism of copper. In some studies very sensitive relationships have been shown, whereas no antagonism was shown in other studies. Solomons [133] has reviewed the difficulties of assessing zinc and copper nutriture in humans, and these may account for some of the discrepancies. Effects on cholesterol metabolism appear to be more sensitive than hematological changes, at least during the periods investigated.

Mertz [134] reviewed the relationship between trace elements and atherosclerosis. He concluded that the major link between trace elements and cardiovascular disease was the impact of various trace elements on risk factors such as disorders of blood lipids, blood pressure, coagulation, glucose tolerance, and circulating insulin. Klevay [135] has recently reviewed the role of copper, zinc, and other chemical elements in ischemic heart disease, as well as several factors that can influence copper status.

The low intakes of copper cited in Sec. 4.1 agree with other limited studies in several parts of the world [134]. Similarly, calcium intake is low in many individuals. The interactions between zinc and calcium and other metals need to be explored further in experimental animals and humans. This should include the variety of factors that can influence the absorption and metabolism of zinc and other elements. Animal studies modeled more closely on human diets and other factors, such as vitamin-mineral supplement and drug use, are needed.

The Joint Food and Agriculture Organization/World Health Organization Expert Committee on Food Additives [136] has proposed a provisional maximum daily tolerable intake for zinc of 1 mg/kg

body weight. At present this seems like a reasonable estimate;
however, it may need to be revised downward with additional research
data.

5. GENERAL CONCLUSIONS

Low dietary intake of essential nutrients and resultant low nutritional status place an individual at increased risk from exposure
to a toxic level of a metal. The risk is greater if the low intake
includes nutrients antagonized by the toxic level of the metal. The
recommended approach to attain adequate nutrition is by consumption
of an adequate diet that includes all types of foods. Consideration
of health factors other than metal exposure per se includes recommendations of low fat intake (particularly saturated fats and cholesterol) and inclusion of foods that provide dietary fiber. Balance
is important for the mineral component of the diet. Consumption of
a single mineral supplement or a vitamin such as ascorbic acid, which
influences mineral absorption, can cause an imbalance more readily
than can consumption of a multiple vitamin-mineral supplement. Use
of the latter at levels not exceeding the RDA has much less potential
for creating undesirable effects on the nutrition of the subject.
Decisions regarding the supplementation of a metal-exposed population
require evaluation by health professionals of dietary intake, nutritional status, metal exposure level, and adverse health effects due
to diet and/or metal before making recommendations for the use of
dietary supplements and other treatment.

REFERENCES

1. National Research Council, *Mineral Tolerance of Domestic Ani-
    mals,* National Academy of Sciences, Washington, D.C., 1980.

2. Anonymous, *Nutrient Intakes: Individuals in 48 States, Year
    1977-78,* Nationwide Food Consumption Survey 1977-78, Report
    No. 1-2, U.S. Department of Agriculture, Hyattsville, Maryland,
    1984.

3. J. A. T. Pennington, D. B. Wilson, B. F. Harland, and J. E. Vanderveen, *J. Am. Diet. Assoc.*, *84*, 771 (1984).

4. National Research Council, *Recommended Dietary Allowances*, National Academy of Sciences, Washington, D.C., 1980.

5. S. M. Pilch and F. R. Senti (eds.), *Assessment of the Iron Nutritional Status of the U.S. Population Based on Data Collected in the Second National Health and Nutrition Examination Survey, 1976-1980,* Life Sciences Research Office, Bethesda, Maryland, 1984.

6. F. R. Senti and S. M. Pilch (eds.), *Assessment of the Folate Nutritional Status of the U.S. Population Based on Data Collected in the Second National Health and Nutrition Examination Survey, 1976-1980,* Life Sciences Research Office, Bethesda, Maryland, 1984.

7. S. M. Pilch and F. R. Senti (eds.), *Assessment of the Zinc Nutritional Status of the U.S. Population Based on Data Collected in the Second National Health and Nutrition Examination Survey, 1976-1980,* Life Sciences Research Office, Bethesda, Maryland, 1984.

8. J. T. McDonald, M. L. Stewart, and R. E. Schucker, *Fed. Proc.*, *42*, 530 (1983).

9. H. E. Ganther, *Ann. N.Y. Acad. Sci.*, *355*, 212 (1980).

10. P. D. Whanger, J. W. Ridlington, and C. L. Holcomb, *Ann. N.Y. Acad. Sci.*, *355*, 333 (1980).

11. J. Parizek, J. Kalouskova, J. Benes, and L. Pavlik, *Ann. N.Y. Acad. Sci.*, *355*, 347 (1980).

12. R. J. Shamberger, *Biochemistry of Selenium*, Plenum Press, New York, 1983.

13. K. Schwarz and C. M. Foltz, *J. Am. Chem. Soc.*, *79*, 3292 (1957).

14. E. J. Underwood, *Trace Elements in Human and Animal Nutrition*, 4th ed., Academic Press, New York, 1977.

15. T. C. Stadtman, *Annu. Rev. Biochem.*, *49*, 93 (1980).

16. J. T. Rotruck, A. L. Pope, H. E. Ganther, and W. G. Hoekstra, *J. Nutr.*, *102*, 689 (1972).

17. J. T. Rotruck, A. L. Pope, H. E. Ganther, A. B. Swanson, D. G. Hafeman, and W. G. Hoekstra, *Science*, *179*, 588 (1973).

18. L. Flohe, W. A. Gunzler, and H. H. Schock, *FEBS Lett.*, *32*, 132 (1972).

19. S. H. Oh, H. E. Ganther, and W. G. Hoekstra, *Biochemistry*, *13*, 1825 (1974).

20. Y. C. Awasthi, E. Beutler, and S. K. Srivastava, *J. Biol. Chem.*, *250*, 5144 (1975).

21. R. A. Lawrence and R. F. Burk, *Biochem. Biophys. Res. Commun.*, *71*, 952 (1976).

22. N. D. Pedersen, P. D. Whanger, P. H. Weswig, and O. H. Muth, *Bioinorg. Chem.*, *2*, 33 (1972).

23. A. I. Calvin, *J. Exp. Zool.*, *204*, 445 (1978).

24. Food and Drug Administration, *Fed. Reg.*, *39*, 1355 (1974).

25. Food and Drug Administration, *Fed. Reg.*, *44*, 5392 (1979).

26. G. F. Combs, Jr., and S. B. Combs, *Annu. Rev. Nutr.*, *4*, 257 (1984).

27. S. O. Welsh, J. M. Holden, W. R. Wolf, and O. A. Levander, *J. Am. Diet. Assoc.*, *79*, 277 (1981).

28. V. C. Morris and O. A. Levander, *J. Nutr.*, *100*, 1383 (1970).

29. T. Westermarck, P. Rannu, M. Kirjarinta, and L. Lappalainen, *Acta Pharmacol. Toxicol.*, *40*, 465 (1977).

30. P. Varo and P. Koivistoinen, *Int. J. Vitam. Nutr. Res.*, *51*, 79 (1981).

31. M. F. Robinson and C. D. Thomson, in *Selenium in Biology and Medicine* (J. E. Spallholz, J. L. Martin, and H. E. Ganther, eds.), AVI, Westport, Connecticut, 1981, p. 24 ff.

32. Keshan Disease Research Group, *Chin. Med. J.*, *92*, 471 (1979).

33. X. Chen, G. Yang, J. Chen, X. Chen, Z. Wen, and K. Ge, *Biol. Trace Element Res.*, *2*, 91 (1980).

34. X. Chen, X. Chen, G. Q. Yang, Z. Wen, J. Chen, and K. Ge, in *Selenium in Biology and Medicine* (J. E. Spallholz, J. L. Martin, and H. E. Ganther, eds.), AVI, Westport, Connecticut, 1981, p. 171 ff.

35. G. Bo-Qi, *Chin. Med. J.*, *96*, 251 (1983).

36. I. Lombeck, K. Kasperek, H. D. Harbisch, K. Becker, E. Shumann, W. Schroter, L. E. Feinendegen, and H. J. Bremer, *Eur. J. Pediatr.*, *128*, 213 (1978).

37. A. M. van Rij, C. D. Thomson, J. M. McKenzie, and M. F. Robinson, *Am. J. Clin. Nutr.*, *32*, 2076 (1979).

38. O. A. Levander, B. Sutherland, V. C. Morris, and J. C. King, *Am. J. Clin. Nutr.*, *34*, 2662 (1981).

39. H. E. Ganther, *Adv. Nutr. Res.*, *2*, 107 (1979).

40. K. P. McConnell and G. J. Cho, *Am. J. Physiol.*, *213*, 150 (1967).

41. C. D. Thomson and R. D. H. Stewart, *Br. J. Nutr.*, *30*, 139 (1973).

42. O. E. Olson, E. J. Novacek, E. I. Whitehead, and I. S. Palmer, *Phytochemistry*, *9*, 1181 (1970).

43. K. P. McConnell and J. L. Hoffman, *FEBS Lett.*, *24*, 60 (1972).

44. R. A. Sunde, G. E. Gutzke, and W. G. Hoekstra, *J. Nutr.*, *111*, 76 (1981).

45. R. A. Sunde and W. G. Hoekstra, *Biochem. Biophys. Res. Commun.*, *93*, 1181 (1980).

46. L. M. Cummins and J. L. Martin, *Biochemistry*, *6*, 3162 (1967).

47. A. T. Nahapetian, M. Janghorbani, and V. R. Young, *J. Nutr.*, *113*, 401 (1983).

48. B. Venugopal and T. D. Luckey, *Metal Toxicity in Mammals*, Vol. 2, Plenum Press, New York, 1978, p. 235 ff.

49. K. W. Franke, *J. Nutr.*, *9*, 596 (1934).

50. G. Yang, S. Wang, R. Zhou, and S. Sun, *Am. J. Clin. Nutr.*, *37*, 872 (1983).

51. F. Berschneider, M. Hess, K. Neuffer, and S. Willer, *Arch. Exp. Veterinaermed.*, *30*, 627 (1976).

52. D. M. Hadjimarkos, *Nutr. Rep. Int.*, *1*, 175 (1970).

53. A. L. Moxon and M. Rhian, *Physiol. Rev.*, *23*, 305 (1943).

54. I. Rosenfeld and O. A. Beath, *Selenium*, Academic Press, New York, 1964.

55. J. R. Harr and O. H. Muth, *Clin. Toxicol.*, *5*, 175 (1972).

56. I. S. Palmer and O. E. Olson, *J. Nutr.*, *104*, 306 (1974).

57. I. Rosenfeld and O. A. Beath, *Proc. Soc. Exp. Biol. Med.*, *87*, 295 (1954).

58. I. Rosenfeld and O. A. Beath, *J. Agric. Res.*, *75*, 93 (1947).

59. R. C. Wahlstrom and O. E. Olson, *J. Anim. Sci.*, *18*, 141 (1959).

60. C. A. Dinkel, J. A. Minyard, and D. E. Ray, *J. Anim. Sci.*, *22*, 1043 (1963).

61. R. M. Jacobs, unpublished observations, 1982.

62. G. N. Schrauzer, personal communication, 1982.

63. J. L. Martin and J. A. Hurlburt, *Phosphorus Sulfur*, *1*, 295 (1976).

64. J. L. Hoffman, *Arch. Biochem. Biophys.*, *179*, 136 (1977).

65. O. A. Levander and V. C. Morris, *J. Nutr.*, *100*, 1111 (1970).

66. M. I. Smith and E. F. Stohlman, *J. Pharmacol. Exp. Ther.*, *70*, 270 (1940).

67. A. W. Halverson and K. J. Monty, *J. Nutr.*, *70*, 100 (1960).

68. A. W. Halverson, C. M. Henderick, and O. E. Olson, *J. Nutr.*, *56*, 51 (1955).

69. I. S. Palmer, O. E. Olson, A. W. Halverson, R. Miller, and C. Smith, *J. Nutr.*, *110*, 145 (1980).

70. J. R. Glover, *Ind. Med. Surg.*, *39*, 50 (1970).

71. G. F. Hunter, in *The Diseases of Occupation*, English University Press, London, 1984, p. 474 ff.

72. D. S. E. Robertson, *Lancet*, *1*, 518 (1970).

73. D. J. Diskin, C. L. Tomasso, J. C. Alper, M. L. Glaser, and S. E. Fliegel, *Arch. Intern. Med.*, *139*, 824 (1979).

74. R. F. Carter, *Med. J. Aust.*, *1*, 525 (1966).

75. I. D. S. Civil and M. J. McDonald, *N.Z. Med. J.*, *87*, 354 (1978).

76. H. Sakurai and K. Tsuchiya, *Environ. Physiol. Biochem.*, *5*, 107 (1975).

77. G. N. Schrauzer and D. A. White, *Bioinorg. Chem.*, *8*, 303 (1978).

78. W. G. Jaffe, D. M. Raphael, M. C. de Mondragon, and M. A. Cuevas, *Arch. Latinoam. Nutr.*, *22*, 595 (1972).

79. M. C. de Mondragon and W. G. Jaffe, *Arch. Latinoam. Nutr.*, *26*, 341 (1976).

80. M. I. Smith and B. B. Westfall, *Public Health Rep.*, *52*, 1375 (1937).

81. R. E. Lemly and L. Merryman, *Lancet*, *61*, 435 (1941).

82. Anonymous, Centers for Disease Control, *Morbid. Mortal. Week. Rep.*, *33*, 157 (1984).

83. K. Helzlsouer, personal communication, 1984.

84. Anonymous, *FDA Drug Bull.*, *14*, 19 (1984).

85. O. A. Levander, G. Alfthan, H. Arvilommi, C. G. Gref, J. K. Huttumen, M. Kataja, P. Koivistoinen, and J. Pickkarainen, *Am. J. Clin. Nutr.*, *37*, 887 (1983).

86. C. D. Thomson, M. F. Robinson, D. R. Campbell, and H. M. Rea, *Am. J. Clin. Nutr.*, *36*, 24 (1982).

87. Anonymous, unpublished GRAS limitation, Food and Drug Administration, Washington, D.C.

88. O. A. Levander, D. P. DeLoach, V. C. Morris, and P. B. Moser, *J. Nutr.*, *113*, 55 (1983).

89. B. L. Vallee and A. Galdes, *Adv. Enzymol.*, *56*, 283 (1984).

90. B. L. Vallee, *J. Inher. Metab. Dis.*, *6*, Suppl., *1*, 31 (1983).

91. W. J. Bettger and B. L. O'Dell, *Life Sci.*, *28*, 1425 (1981).

92. A. S. Prasad, I. E. Dreosti, and B. S. Hetzel (eds.), *Clinical Applications of Recent Advances in Zinc Metabolism*, Alan R. Liss, New York, 1982.

93. A. S. Prasad, A. O. Çavdar, G. J. Brewer, and P. J. Aggett, *Zinc Deficiency in Human Subjects*, Alan R. Liss, New York, 1983.

94. N. W. Solomons, *Am. J. Clin. Nutr., 35,* 1048 (1982).

95. G. E. Inglett (ed.), *Nutritional Bioavailability of Zinc,* American Chemical Society, Washington, D.C., 1983.

96. D. R. Grant-Frost and E. J. Underwood, *Aust. J. Exp. Biol. Med. Sci., 36,* 339 (1958).

97. C. H. Hill and G. Matrone, in *Proceedings of the Twelfth World's Poultry Congress* (A. B. Welcom, ed.), World's Poultry Science Association, Clun, Shropshire, England, 1962, p. 219 ff.

98. R. P. Hamilton, M. R. S. Fox, B. E. Fry, Jr., A. O. L. Jones, and R. M. Jacobs, *J. Food Sci., 44,* 738 (1979).

99. J. K. Campbell and C. F. Mills, *Proc. Nutr. Soc., 33,* 15A (1974).

100. R. Van Reen, *Arch. Biochem. Biophys., 46,* 337 (1953).

101. C. T. Settlemire and G. Matrone, *J. Nutr., 92,* 159 (1967).

102. R. P. Hamilton, M. R. S. Fox, S.-H. Tao, and B. E. Fry, Jr., *Nutr. Res., 1,* 589 (1981).

103. A. C. Magee and G. Matrone, *J. Nutr., 72,* 233 (1960).

104. D. H. Cox and D. L. Harris, *J. Nutr., 70,* 514 (1960).

105. C. H. Hill and G. Matrone, *Fed. Proc., 29,* 1474 (1970).

106. C. H. Hill, G. Matrone, W. L. Payne, and C. W. Barber, *J. Nutr., 80,* 227 (1963).

107. C. R. Bunn and G. Matrone, *J. Nutr., 90,* 395 (1966).

108. K. Guggenheim, *Blood, 23,* 786 (1964).

109. D. G. Thawley, R. A. Willoughby, B. J. McSherry, G. K. MacLeod, K. H. MacKay, and W. R. Mitchell, *Environ. Res., 14,* 463 (1977).

110. V. Sandisivan, *Biochem. J., 48,* 527 (1951).

111. A. K. Stewart and A. C. Magee, *J. Nutr., 82,* 287 (1964).

112. R. A. Willoughby, E. MacDonald, B. J. McSherry, and J. Brown, *Vet. Rec., 91,* 382 (1972).

113. M. R. S. Fox, R. P. Hamilton, A. O. L. Jones, B. E. Fry, Jr., and R. M. Jacobs, *Abstracts of the XI International Congress on Nutrition,* 1978, p. 140.

114. I. Bremner and N. T. Davies, *Biochem. J., 149,* 733 (1975).

115. D. R. Van Campen and P. U. Scaife, *J. Nutr., 91,* 473 (1967).

116. A. C. Hall, B. W. Young, and I. Bremner, *J. Inorg. Biochem., 11,* 57 (1979).

117. L. M. Klevay, *Am. J. Clin. Nutr., 28,* 764 (1975).

118. K. G. D. Allen and L. M. Klevay, *Nutr. Rep. Int., 22,* 295 (1980).

119. L. M. Klevay, *Abstracts of the Symposium on Trace Elements in Man and Animals,* p. 28 (1984).

120. H. Spencer, L. Kramer, C. Norris, and D. Osis, *Am. J. Clin. Nutr., 35,* 816 (1982).

121. J. L. Greger, S. C. Zahie, R. P. Abernathy, O. H. Bennett, and J. Huffmann, *J. Nutr., 108,* 1449 (1978).

122. D. M. Burke, F. J. DeMicco, L. J. Taper, and S. J. Ritchey, *J. Gerontol., 36,* 558 (1980).

123. H. H. Sandstead, *Am. J. Clin. Nutr., 35,* 809 (1982).

124. D. M. Gallery, J. Blomfield, and S. R. Dixon, *Br. Med. J., 4,* 331 (1972).

125. J. J. B. Petrie and P. G. Row, *Lancet, 1,* 1178 (1977).

126. K. M. Hambidge, P. A. Walravens, K. H. Nelder, and N. A. Daugherty, in *Trace Element Metabolism in Man and Animals—3* (M. Kirchgessner, ed.), University of Munich, Freising-Weihenstephan, 1978, p. 413 ff.

127. K. G. Porter, D. McMaster, M. E. Elmes, and A. H. G. Love, *Lancet, 2,* 774 (1977).

128. A. S. Prasad, G. J. Brewer, E. B. Schoomaker, and P. Rabbani, *J. Am. Med. Assoc., 240,* 2166 (1978).

129. M. Abdulla, *Lancet, 1,* 616 (1979).

130. J. H. Freeland-Graves, W.-H. Han, B. J. Frieden, and R. A. L. Shorey, *Nutr. Rep. Int., 22,* 285 (1980).

131. P. L. Hooper, L. Visconti, P. J. Garry, and G. E. Johnson, *J. Am. Med. Assoc., 244,* 1960 (1980).

132. L. M. Klevay, L. Inman, L. K. Johnson, M. Lawler, J. R. Mahalko, D. B. Milne, H. C. Lukaski, W. Bolonchuk, and H. H. Sandstead, *Metabolism, 33,* 1112 (1984).

133. N. W. Solomons, *Am. J. Clin. Nutr., 32,* 856 (1979).

134. W. Mertz, *Fed. Proc., 41,* 2807 (1982).

135. L. M. Klevay, in *Metabolism of Trace Elements in Man,* Vol. 1 (O. W. Rennert and W.-Y. Chan, eds.), CRC Press, Boca Raton, Florida, 1984, p. 129 ff.

136. Joint FAO/WHO Expert Committee on Food Additives, *WHO Tech. Rep. Ser., 683,* 32 (1982).

8

Chromosome Damage in Individuals Exposed to Heavy Metals

Alain Léonard
Mammalian Genetics Laboratory
Department of Radiobiology
Center for the Study of Nuclear Energy
B-2400 Mol, Belgium
and
Teratogenicity and Mutagenicity Unit
University of Louvain
Brussels, Belgium

1. INTRODUCTION

Professional or accidental exposure to ionizing radiation is currently monitored on the basis of chromosome aberrations induced in peripheral blood lymphocytes, and the dose is estimated by comparison with dose-response relationships obtained in vitro [1-4]. On the assumption that agents that increase the yield of chromosomal aberrations are able to produce point mutations, this technique also represents one of the easiest methods allowing a direct estimation of the genetic damages induced by environmental mutagens in human somatic cells. The possibility of using this system for exposure to mutagens other than ionizing radiations was therefore very attractive, and many papers have been published reporting an increase of structural chromosome aberrations in people exposed to chemicals in general and to metals in particular [5-8].

2. PRODUCTION OF CYTOGENETIC CHANGES IN PERIPHERAL BLOOD LYMPHOCYTES

2.1. General Characteristics of the Peripheral Blood Lymphocyte System

About 90% of the peripheral blood lymphocytes are of the recirculating type, remaining in the bloodstream for less than 1 hr. They then migrate to the lymphatic tissues to reappear in the blood hours or days later [9,10]. More than 99% of the circulating lymphocytes are in the G_0 or G_1 stage of the cell cycle and have to be stimulated by a mitogenic agent, generally phytohemagglutinin, to replicate their DNA and to enter in division. The studies performed on irradiated persons have shown [11,12] that the majority (90%) of human lymphocytes have a life span exceeding 3 years, some surviving more than 20 years. Since lymphocytes thus accumulate genetic damages, they allow one to monitor the effects of acute as well as chronic or repeated exposures to clastogenic agents.

2.2. Structural Chromosome Anomalies Produced by Clastogens

Chromosome type and chromatid type are the most frequent structural aberrations which can be cytologically distinguished at metaphase after exposure to clastogenic agents. Chromosome-type aberrations occur at the same locus of both chromatids, whereas chromatid-type aberrations usually involve only one chromatid of the metaphase chromosomes. Chromosome-type anomalies include terminal deletions, minute or interstitial deletions, acentric rings, centric rings, inversions, reciprocal translocations, and polycentric aberrations. The most common chromatid-type aberrations are chromatid and iso-chromatid gaps, chromatid breaks, and chromatid interchanges (Figs. 1-5). The chromatid and isochromatid gaps (achromatic lesions) are generally considered unreliable indicators of real damage to the genetic material [5,13-15].

Some chemicals, especially those that inhibit the synthesis of DNA, usually induce a high frequency of such achromatic lesions when cells are treated in the S or G_2 stage. As pointed out by Evans [13], it is probable that gaps involve a mixture of phenomena, some representing true lesions to genetic material. Their scoring is, however, extremely subjective, and their frequency is sometimes very high in control cells and may be increased by differences in technique. In order to avoid a selective elimination of cells carrying unstable aberrations, the dividing cells are normally analyzed in the metaphase stage of the first mitosis (M_1) following their induction.

Comparative studies have shown that the structural changes produced by clastogenic chemicals are similar to those produced by

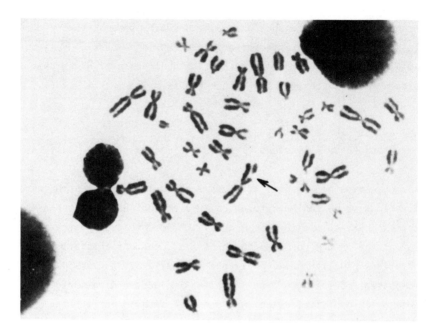

FIG. 1. Metaphase showing a chromatid gap.

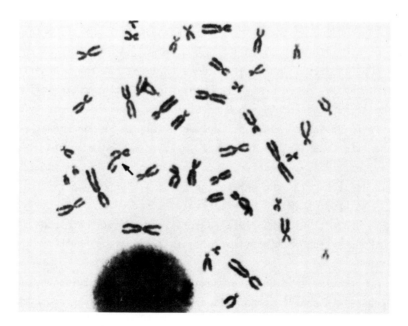

FIG. 2. Metaphase showing a chromatid break.

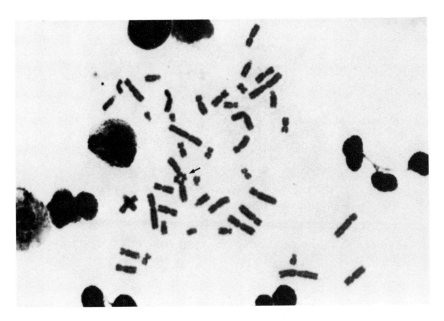

FIG. 3. Metaphase with a chromatid exchange.

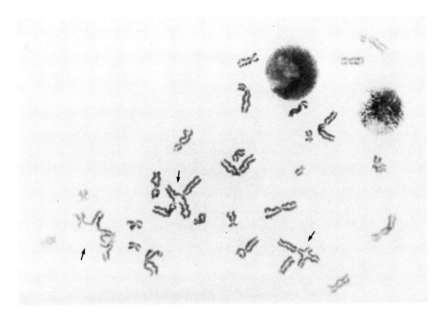

FIG. 4. Metaphase with multiple chromatid-type aberrations.

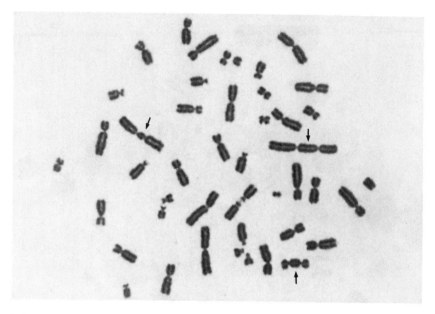

FIG. 5. Metaphase with dicentric chromosomes.

ionizing radiations, the frequency of the different types of anomalies depending both on the stage of the cell cycle in which the
exposure occurs and on the type of clastogenic agent. Ionizing
radiations induce chromosome-type aberrations if the cells are
exposed in G_1, chromatid-type aberrations if the cells are exposed
in G_2, and a mixture of both types if exposure occurs in the S
phase. In contrast to ionizing radiations, chemical mutagens have
a wide spectrum of mechanisms by which they exert their mutagenic
action. Most produce structural aberrations by a so-called delayed
effect and, after an intervening round of DNA replication, give rise
to chromatid-type aberrations. Some chemicals, however, mimic ionizing radiations and produce chromosome-type structural aberrations
within a short time after exposure. Chemicals often are either S
dependent or S independent, but some compounds apparently possess
both capabilities.

The micronucleus test on peripheral blood lymphocytes may
represent an interesting alternative to the observation of metaphase

chromosomes [16-21] and has been employed occasionally for biomonitoring people exposed to heavy metals [22].

2.3. Sister Chromatid Exchanges

The use of 5-bromodeoxyuridine [23] has greatly simplified the detection of sister chromatid exchanges (SCEs) and explains the proliferation, during the last decade, of articles dealing with the induction of such cytogenetic changes by various physical and chemical agents. This technique, which gives better resolution than autoradiography, requires the exposure of the cells to 5-bromodeoxyuridine during two consecutive replications. The 5-bromodeoxyuridine is incorporated into both strands of one chromatid and only one strand of its sister chromatid. Subsequent treatment of the M_2 chromosomes with a combination of the fluorochrome Hoechst 33528 and Giemsa results in a differential staining of the two sister chromatids, allowing easy detection of the exchanges (Fig. 6). In vitro observations have

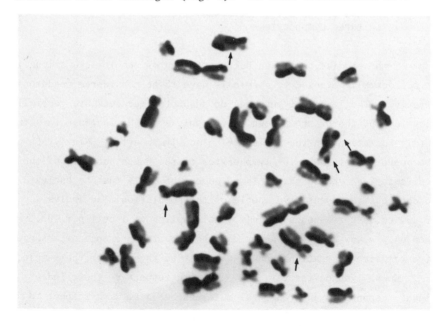

FIG. 6. M_2 metaphase with sister chromatid exchanges.

shown that, in contrast to the S-independent agents, the S-dependent clastogenic agents are generally efficient in inducing SCEs [24]. No satisfactory mechanism has been proposed up to now for the production of SCEs, and their biological significance is still unknown. Nevertheless, since the effects of many mutagenic agents can be readily detected at levels which cause almost no morphological chromosome damage, the induction of SCEs appears to be one of the most sensitive mammalian systems for detecting in vitro the mutagenic potential of chemicals.

The first results obtained on patients treated with cytostatic agents [25-29] or after exposure to other toxic agents [30,31] suggested that the scoring of SCEs in peripheral blood lymphocytes could represent an interesting method for detecting significant exposures of individuals to mutagens. Up to now, however, the method has not been proven to be a good monitor of human exposure for the majority of compounds analyzed [32].

2.4. Drawbacks of the Cytogenetic Analysis in Cultured Lymphocytes

Many of the studies that have been performed on people occupationally or accidentally exposed to chemicals have to be considered nondemonstrative. The levels of mutagens or suspected mutagens are generally very low and therefore one can expect only a slight increase in structural aberrations in the peripheral blood lymphocytes. Nearly 50% of the approximately $500 \cdot 10^9$ lymphocytes of the human body are of the recirculating type so that a large number of cells must be analyzed to obtain a representative sample of the population: The analysis of 250 cells corresponds roughly to a sampling of 1 cell per 10^9. When only a few cells are analyzed, the reported presence or absence of chromosome aberrations can be a result of the size of the sample.

Human lymphocytes vary not only with respect to their individual response to environmental mutagens but also with respect to their mitogenic activation. As early as 48 hr after culture initiation, the culture time generally used, large numbers of second mitoses may already be present, leading to erroneous estimation of the yield of aberrations produced. At each cell generation about 50% of the unstable chromosome aberrations are eliminated [33-36], and chromosome aberrations can develop from original chromatid-type aberrations as a result of the second division of cells [13,15]. Furthermore, studies on normal and irradiated human blood cells have shown [13] that the incidence of gaps and chromatid aberrations in lymphocyte chromosomes increases as cells are allowed to proceed through two or more cycles in culture.

Problems can also arise because of the fact that human populations are often exposed to a mixture of potential mutagens. Therefore any study should take into account occupational history, age, previous or current exposure to drugs, toxic substances, ionizing radiations, and viral infections. Such interferences probably explain the frequent articles reporting an increase of chromosome-type aberrations such as centric rings or polycentric chromosomes in people exposed to S-dependent chemicals.

Proper controls are indispensable to detect an increase in the yield of structural changes above the normal aberration yield. Such control individuals should have a closely similar environmental background and be matched by sex and age with the exposed persons [15]. Ideally, the best method would be to use cells from the same person before exposure, but this is rarely possible. Finally, the blood samples from control and exposed groups should be processed together in order to minimize differences due to methodological factors [15].

3. RESULTS OF THE CYTOGENETIC MONITORING PERFORMED ON PERSONS EXPOSED TO HEAVY METALS

3.1. Arsenic

Several cytogenetic studies report conflicting results on peripheral blood lymphocytes from people exposed to arsenic. They include observations on patients treated with arsenic [37-39], wine growers contaminated by arsenic-containing pesticides [40], and workers

TABLE 1

Results of Observations on People Exposed to Arsenic

Test	Type of exposure	Results	References
Structural chromosome aberrations	Medical treatment	+	37
	Medical treatment	+	39
	Medical treatment	-	38
	Pesticides	+	40
	Smelter	+	41-44
SCE	Medical treatment	+	38
	Medical treatment	-	39

exposed to airborne arsenic at the Rönnskär copper smelter in Northern Sweden (Table 1) [41-44].

Arsenicals have been shown to be powerful clastogens in experiments performed in vitro on different mammalian cell types [45], and slightly positive results have been obtained in vivo in laboratory animals [46]. It is, however, doubtful whether arsenic possesses such clastogenic properties in vivo in human somatic cells. All studies have been performed on lymphocytes cultured for 72 hr, with a small number of cells (generally 100), and gaps contributed greatly to the increase of structural aberrations. In the smelters the yield of structural aberrations (gaps, chromatid and chromosome breaks) was significantly increased compared to the controls, but the correlation between the frequency of aberrations and arsenic exposure was rather poor except for chromosome-type aberrations. Since in such plants the workers are generally exposed to a mixture of possible mutagens, for example, lead, selenium, and copper, the authors suggested that the role of arsenic was only to inhibit the repair of damages produced by other environmental agents. Observations on patients treated with arsenic could clarify the situation, because doses are generally higher and fewer possibilities of interference with other mutagenic agents exist. Petres et al. [37] and Nordenson et al. [39] reported

a significant enhancement of aberrations in such persons, but this
was not confirmed by Burgdorf et al. [38], in spite of the fact that
some patients received, in addition to arsenic and for therapeutic
purposes, local irradiation doses as high as 5000 rads!

A similar apparent contradiction exists with respect to the
ability of arsenic to increase the yield of SCEs in vivo in human
peripheral blood lymphocytes: Burgdorf et al. [38] found a threefold
increase of SCEs in six patients treated with arsenic, whereas the
SCE rate was found normal in both arsenic-treated and untreated
patients examined by Nordenson et al. [39]. It should be recalled
that an increase of SCEs has been obtained in vitro with all the
arsenic salts tested up to now [45].

3.2. Cadmium

The majority of the assays performed in vitro and in vivo to evaluate
the clastogenic properties of cadmium salts in mammalian cells yielded
negative results [47], and this was confirmed by Bui et al. [48] on
four female itai-itai patients and five male workers from an alkaline
battery factory and by O'Riordan et al. [49] on a population of 40
men employed in the manufacture of cadmium pigments (Table 2). In

TABLE 2

Results of Observations on People Exposed to Cadmium

Test	Type of exposure	Results	References
Structural chromosome aberrations	Itai-itai disease	+	55,56
	Itai-itai disease	-	48
	Pigment industry	-	49
	Zinc smelter	+	51,52
	Zinc smelter	+	53,54
	Battery manufactory	-	48

both studies none of the individuals had received radiotherapy or drugs known to produce chromosome damage. The cadmium blood level at the time of the cytogenetic observations were 1.95 µg/100 ml versus 0.2 µg/100 ml in the controls in the group studied by O'Riordan et al. [49] and, in the study of Bui et al. [48], ranged from 24.7 to 61.0 ng/g and from 15.5 to 28.8 ng/g for the industrial workers and the itai-itai patients, respectively.

These negative findings contrast markedly, however, with the positive results reported by several authors on workers employed in zinc smelting plants [50-54] or on Japanese itai-itai patients [55, 56].

In these studies on industrial workers, the cadmium content of blood did not differ greatly from the values reported by O'Riordan et al. [49], and no correlation was found with the degree of exposure. Furthermore, it should be pointed out that persons working in such industries are exposed to a mixture of different metals, for example, cadmium, lead, and zinc. Therefore there is the possibility that synergistic effects of chronic exposure to several metal compounds influence both the type and yield of structural aberrations. Itai-itai disease represents a particularly severe manifestation of chronic cadmium poisoning and was observed in several hundreds of middle-aged women in a Japanese village near Toyama. The name itai-itai, or "ouch-ouch," disease originates from the excruciating pain due to multiple fractures arising from osteomalacia. This intoxication resulted from the ingestion of rice irrigated with water from the Yintsu River, which was contaminated with cadmium, lead, and zinc from a zinc mine, resulting in an increased cadmium content in crops in that area. In addition to possible synergism, the increase in the yield of structural aberrations also could result from the fact that most of those patients received heavy radiodiagnostic exposures and various drugs.

3.3. Chromium

Hexavalent chromium has been shown to display clastogenic properties
in vitro in mammalian cells and in vivo in laboratory animals, whereas
trivalent chromium appeared generally inactive [50]. The positive
results reported by Bigaliev et al. [57] and Azhajev [58] on workers
engaged in chrome production have not been confirmed, however, by
Husgafvel-Pursiainen et al. [59] and Littorin et al. [22] for stainless steel manual metal arc welders exposed to fumes containing
chromium(III) and nickel (Table 3). In the study of Littorin et al.
[22] the mean exposure time was 19 years, the air chromium level
averaged 81 $\mu g/m^3$ at the time of exposure, and the urinary chromium
was 47 $\mu mol/mol$ creatinine, but, in spite of the high level of exposure, the yields of structural chromosome aberrations and of micronuclei or sister chromatid exchanges were not significantly different
from the values found in the matched controls. In the three studies
structural chromosome aberrations were observed after a cultivation
time of 72 hr and the number of cells examined was relatively low.

TABLE 3

Results of Observations on People Exposed to Chromium

Test	Type of exposure	Results	References
Structural chromosome aberrations	Chromium plant	+	57
	Stainless steel welder	-	22
	Stainless steel welder	-	59
SCE	Chromium plant	-	58
	Stainless steel welder	-	59
	Stainless steel welder	-	22

3.4. Lead

The numerous cytogenetic studies performed on peripheral blood
lymphocytes demonstrate the confusing feature of the ability of lead
to produce chromosome aberrations in the somatic cells of people
professionally or accidentally exposed (Table 4). Positive reports
have been published for workers from industry [54,60-70], children
living in contaminated area [69], and bus drivers [71], but equally
convincing negative results have been presented by other researchers
for people exposed under comparable conditions, for example, industrial workers [72-78], children living in contaminated areas [79],
and policemen [80,81]. The qualitative nature of the results can
apparently not be explained by methodological peculiarities. In the
studies reporting positive results, as well as in those with negative
findings, the culture time varied from 48 to 72 hr. Furthermore, no
relation was detectable between the level of exposure and the results
obtained: Workers from the zinc industry analyzed by Bauchinger et
al. [54] showed an increase of structural aberrations at a blood lead
level averaging 19.29 µg/100 ml, whereas O'Riordan and Evans [75]
failed to detect an enhancement of chromosome aberrations above control values in a population of men exposed to oxides of lead in a
shipbreaking yard with the blood lead level occasionally exceeding
120 µg/100 ml. It is of interest, also, to note that some authors
[83] were unable to confirm later, positive results obtained previously on the same population [62].

The results of experimental studies performed under well-
controlled conditions on laboratory animals strongly suggest that
lead is devoid of clastogenic properties. Jacquet et al. [82] as
well as Heddle and Bruce [83] could not observe an increase in micronuclei in mouse bone marrow erythocytes after an intraperitoneal
injection of 50 mg/kg lead acetate and, similarly, no aberrations
were detected in the metaphases of bone marrow cells of mice given
1 g/liter lead in their drinking water for 9 months [84] or kept for
1 month on a diet containing 0.5% lead [85]. Jacquet et al. [82]

TABLE 4

Results of Observations on People Exposed to Lead

Test	Type of exposure	Results	References
Structural chromosome aberrations	Lead oxide manufacture	+	60,61
		-	72
	Shipbreaking yard	-	75
	Bus drivers	+	71
	Policemen	-	80,81
	Copper smelter	+	62
		-	73
	Lead smelter	-	73
		-	76
		-	77
		+	54
	Lead industry	+	63
		+	64
		+	65-68
		-	78
	Children living in contaminated area	+	69
		-	79
	Lead intoxication	-	89
		±	94
	Volunteers	-	90
SCE	Lead smelter	±	76
	Children living in contaminated area	-	93

detected a slight dose-dependent increase in gaps when 0, 0.25, 0.5, and 1% doses of dietary lead were given to mice for 1 month, and Deknudt et al. [86] reported an increase in gaps and fragments in cynomolgus monkeys (*Macaca irus*) receiving 1.5, 6, and 15 mg of lead acetate 6 days per week for 16 months, but no fragments were observed in a follow-up of the same material [87]. Muro and Goyer [88] observed gaps and chromatid aberrations in cultured bone marrow cells from mice kept on lead for 2 weeks, but the culture time of 4 days was too long to give meaningful results.

While the possibility cannot be excluded that lead is responsible for certain achromatic lesions (gaps) observed in persons professionally or accidentally exposed, it is evident, however, that this metal cannot be considered an S-independent agent. Since in most positive studies [60-62,64,71] the majority of the patients showed excessive levels of both chromosome- and chromatid-type aberrations, whereas the negative results were obtained in cases of heavy intoxication [89,90], it appears likely that the aberrations observed are surviving events of exposure to other clastogenic agents, for example, irradiations given for radiodiagnostic or radiotherapeutic purposes, or result from a simultaneous exposure to a mixture of different chemicals displaying synergistic or additional clastogenic properties. For instance, as pointed out by the authors themselves, the bus drivers studied by Högstedt et al. [71] had been exposed to automobile exhaust fumes containing, besides lead, several powerful mutagens such as polycyclic aromatic hydrocarbons; also, workers in copper smelter [62] and in zinc industry [51,52] are exposed, in addition to lead, to several other metals.

In agreement with the results obtained in vitro with lead acetate [91] or lead chloride [92], no significant increase in the mean values of SCE frequencies was observed by Dalpra et al. [93] in lymphocytes from the peripheral blood cultures of 19 children who lived in a widely contaminated area and showed an increased absorption of lead. Mäki-Paakkanen et al. [76] report, however, a slight increase in SCE frequency among lead-exposed smokers compared to referents who also smoked.

3.5. Mercury

Production of c-mitosis, which probably results from the high affinity of mercury for sulfhydryl groups in the spindle apparatus, is the most noticeable effect of mercury on genetic material.

Observations on plant material and on mammalian cells in vitro suggest that mercury compounds could display clastogenic properties, but the few experiments carried out so far on mammals in vivo have given negative results. Mice treated with methyl mercury acetate showed no excessive numbers of micronuclei [83,95], and no chromosome aberrations were found by Poma et al. [96] in the spermatogonia and bone marrow cells of mice given a single intraperitoneal injection of 2, 4, or 6 mg $HgCl_2$/kg body weight.

The numerous positive findings reported on peripheral blood lymphocytes of people accidentally or professionally exposed to mercury appear, therefore, rather surprising (Table 5). A significant increase of structural aberrations was found by Skerfving et al. [97-99], in persons who had whole-blood mercury concentrations up to 650 µg/g after the consumption of fish contaminated by methyl mercury (1-7 ppm), and by Kato et al. [100], in Minamata residents having accumulated an average of 211 ppm mercury in their hair owing to the contamination of seafood by methyl mercury chloride (up to 50 ppm in fish and 85 ppm in shellfish) discharged from 1953 to 1960 into the bay and river of Minamata by a plastic manufacturing plant. Similar results have been reported in persons exposed to different mercury compounds, namely, metallic mercury, phenyl and ethyl mercury compounds or amalgams [101-104], and phenyl mercury acetate [103]. It seems possible that some of these positive results are due to other

TABLE 5

Results of Observations on People Exposed to Mercury

Test	Type of exposure	Results	References
Structural chromosome aberrations	Contaminated fish	+	97-99
	Minamata disease	+	100
	Mercury industry	+	101
		+	102-104
	Chloroalkali plant	-	105
		-	106

environmental agents or could be explained by some methodological peculiarities:

1. In the study of Skerfving et al. [97-99] the incubation time varied from 48 to 120 hr in the first set of experiments and was 72 hr in the second.

2. Kato et al. [100] analyzed only 50 cells per person and did not provide details on the methodology used and on the control values.

3. Popescu et al. [101] analyzed 4 men exposed to metallic mercury vapors and 18 subjects exposed to organic mercury and reported a not statistically significant difference between exposed groups and controls with respect to chromatid-type aberrations. There was an evident enhancement of chromosome-type aberrations, but the fact that a man showing 3 dicentric chromosomes and 10 acentric fragments in 170 scored metaphases had formerly been poisoned by benzene indicates that at least some of the structural aberrations can be related to previous or simultaneous exposure to other mutagens displaying clastogenic properties.

4. In a recent study Verschaeve et al. [105] were unable to confirm their previous positive findings [102-104] and failed to detect any significant increase in structural chromosome aberrations in peripheral blood lymphocytes of workers exposed to metallic mercury in a chloroalkali plant. Similar negative findings were reported by Mabille et al. [106] on subjects from a plant where mercury is amalgamated with zinc and on other subjects from a chloroalkali plant; Verschaeve et al. [105] themselves suggested that factors other than mercury compounds, for example, less stringent protection measures, more careless work habits, or concomitant exposure to other risks, could be responsible for their previous positive findings.

3.6. Nickel

Some observations have been made by Boysen et al. [107] and Waksvick et al. [108,109] on subjects working in nickel refineries at Falcon-bridge Nikkelverk, Kristiansand (Norway). In this plant the converter matte (50% Ni, 30% Cu, 20% S, and some trace metals) is refined through several processes, including crushing, roasting, smelting, and electrolysis. A total of 100 cells per person were analyzed for the presence of structural aberrations after 48 hr of culture in peripheral blood lymphocytes from nine male subjects exposed to NiO and Ni_3S_2, ten males exposed to $NiCl_2$ and $NiSO_4$, and seven controls. The plasma concentrations of nickel averaged 4.2, 5.2, and 1 µg/liter, respectively. The summary of the results presented in Table 6 reveals a drastic increase in the frequency of gaps among the workers, whereas there is only a slight enhancement of breaks. These positive findings were confirmed later [109] by observing an increase in gaps and breaks in retired workers with

TABLE 6

Results of Observations on People Exposed to Nickel

Exposure	Plasma concentration of nickel (µg/liter)	Number of SCEs per cell	Number of aberrations per 100 cells	
			Gaps	Breaks
NiO,Ni_3S_2	4.2	4.8	11.9	0.9
$NiCl_2$,$NiSO_4$	5.2	4.9	18.3	1.3
Controls	1.0	5.1	3.7	0.6
Retired workers	2.0	6.0	7.6	4.1
Controls	1.0	6.0	5.3	0.5

Source: Refs. 107-109.

more than 25 years of nickel exposure, as is evident also from
changes in the nasal mucosa. The yield of SCEs was not modified
when compared to control values.

Several nickel salts have been shown to be able to enhance
the frequency of SCEs in vitro, but there exists no clear evidence
they have clastogenic properties [110]. Negative findings have
been obtained in laboratory animals with respect to the production
of structural aberrations. For these reasons the role of nickel in
the enhancement of structural aberrations in professionally exposed
people therefore appears questionable.

3.7. Various Metals

An increase in the yield of cells carrying chromosome aberrations
has been observed by Pilinskaya [111], in the peripheral blood
lymphocytes of persons handling zinc dimethyl dithiocarbamate
(Ziram, $C_6H_{12}N_2S_4Zn$), a fungicide also used as a rodent repellent,
and by Kuleshov and Sedova [112], in workers from metallurgy. No
increase in structural chromosome aberrations was reported in the
case of occupational contact with molybdenum [113].

4. GENERAL DISCUSSION

Beside their general cytotoxic action causing denaturation of macromolecules, several metals may enter in competition with an essential
metal needed to stabilize the structure of a biomolecule. On the
basis of their reactions with different ligands, one can distinguish
[114] "hard" ions, for example, Mg and Ca, which bind to phosphate
groups and tend to stabilize the double helix, and "soft" ions, for
example, Hg and Ag, which bind to bases and decrease the stability
of the helix; nickel, cobalt, and zinc are considered as intermediary.
These effects could explain the disturbances of the spiralization of
the chromosomes resulting in a decondensation which has been observed

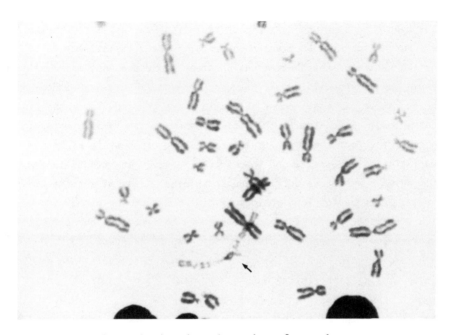

FIG. 7. Metaphase showing decondensation of one chromosome.

in peripheral blood lymphocytes of some workers from the zinc industry
[51,52] and in people exposed to mercury [103] (Fig. 7).

The replacement of Zn^{2+} and Mg^{2+} in metalloenzymes, such as DNA
polymerases, by other cations can cause a decrease in replicational
fidelity [115]. Furthermore, toxic metals can interfere with DNA
repair [116,117]. Some synergistic or additional effects which have
been accounted for to explain the presence of chromosome aberrations
in people intoxicated by heavy metals could be explained by such
actions producing a potentialization of damages produced by other
environmental agents to genetic material.

The results of the experimental studies in vitro on cultured
cells reveal [50] that only arsenic and hexavalent chromium display
obvious clastogenic properties in vitro, whereas no conclusive results
have been obtained with the salts of cadmium, lead, mercury, and
nickel. The conclusions of the in vitro assays have generally been
confirmed in vivo on experimental mammals. Compared to powerful

clastogenic agents such as ionizing radiation, the ability of metals
to induce structural aberrations appears, however, relatively weak,
and the nature of the aberrations observed indicates that metal salts
displaying clastogenic properties cannot be considered as S-independent
agents, since the most common aberrations are gaps and chromatid-type
aberrations. Furthermore, the yield of aberrations is always low in
spite of the fact that in vitro systems imply exposure to high concentrations of the salts and that in the in vivo experiments laboratory animals were treated with acute sublethal doses or submitted to
chronic exposure for long periods of time. Except for cases with very
high intoxication, it is therefore hard to believe that chronic exposures to low doses of weak clastogenic S-dependent agents could enhance
the yield of structural chromosome aberrations detectable in a sample
(50-200 cells) representing only a very small fraction of the cell
population which has been exposed.

Turbagenic compounds interfere with the function of the spindle
apparatus mainly through an inhibition of tubuline polymerization and
induce disorders in chromosome distribution resulting in aneuploid
cells and c-mitosis. Such mitotic poisons cannot be considered true
mutagens because they do not act directly upon the hereditary material
itself, but may contribute to mutagenesis as much as true mutagens.
Turbagenic action appears to be the most evident cytologically detectable effect of exposure of eukaryotic cells to metal salts. Positive
response has indeed been obtained with all metal salts tested up to
now on plant material. The observation of peripheral blood lymphocytes unfortunately does not allow the detection of such effects in
human somatic cells, because the analysis of the cells is performed
in M_1 after their stimulation by phytohemagglutinin.

In the SCE test a positive response has been obtained in vitro
with arsenic, chromium, nickel, and lead salts, whereas mercury gave
negative results. Contradictory results have been obtained on human
peripheral blood lymphocytes after in vivo exposure to arsenic, and
negative results were obtained for chromium, lead, and nickel (Tables
1-6). The SCEs observed after culture of peripheral blood lymphocytes

from persons exposed to chemicals are generally considered as the
results of damage produced in vivo to genetic material [24]. Therefore one cannot exclude the possibility that the level of exposure
in the in vivo studies performed up to now was too low to induce
such changes. The fact that some studies reported for the same
people an absence of enhanced yield of SCEs together with an increase
of structural aberrations [108-110] could be considered as an argument that the chromosome anomalies originate from other environmental
agents. Another possibility, suggested by Hollstein et al. [118], is
that the production of SCEs is determined essentially by the amount
of mutagens present in the culture at the time of DNA synthesis.

5. CONCLUSIONS

The results of the observations performed on peripheral blood lymphocytes of people professionally or accidentally exposed to heavy metals
very often appear contradictory, and methodological features as well
as the levels of intoxication cannot explain the differences obtained
for people exposed under comparable conditions. Furthermore, the
observations performed on peripheral blood lymphocytes frequently
disagree with the properties revealed by the different heavy metals
in the in vitro assays and in the experiments made on laboratory
animals.

Most people are exposed not to a single metal but to a mixture
of environmental mutagenic agents. It is therefore extremely difficult to establish a cause-effect relationship between a low increase
in the yield of chromosome aberrations and one of the many components
encountered in the environment, since these agents interfere with each
other and have synergistic, additional, or inhibitory effects. The
only conclusion that can therefore be drawn from such studies is that
the conditions of work encountered in a determined plant or specific
place have or have not led to an increase, above control values, of
the yield of aberrations observed in the peripheral blood lymphocytes
of persons having also their own habits of life. The value of such

studies would be greatly improved by the adoption of a careful methodology, by the analysis of a large number of cells, and by using appropriate controls. It is also evident that it would be of interest to combine the cytogenetic observations with other assays used for human biomonitoring.

REFERENCES

1. D. C. Lloyd, R. J. Purrott, and E. J. Reeder, *Mutat. Res., 72,* 523 (1980).

2. D. C. Lloyd and R. J. Purrott, *Radiat. Prot. Dosim., 1,* 19 (1981).

3. D. C. Lloyd, *J. Soc. Radiat. Prot., 4,* 5 (1984).

4. M. S. Sasaki, in *Radiation Induced Chromosome Damage in Man* (T. Ishihara and M. S. Sasaki, eds.), Alan R. Liss, New York, 1983, pp. 585-604.

5. G. Obe and B. Beek, in *Chemical Mutagens. Principles and Methods for Their Detection,* Vol. 7 (F. J. Serres and A. Hollaender, eds.), Plenum Press, New York, 1982, pp. 337-400.

6. A. T. Natarajan and G. Obe, in *Mutagenicity. New Horizons in Genetic Toxicology* (J. A. Heddle, ed.), Academic Press, New York, 1982, pp. 171-213.

7. M. Kucerova, in *Mutagenicity. New Horizons in Genetic Toxicology* (J. A. Heddle, ed.), Academic Press, New York, 1982, pp. 241-266.

8. J. G. Brewen and D. G. Stetka, in *Mutagenicity. New Horizons in Genetic Toxicology* (J. A. Heddle, ed.), Academic Press, New York, 1982, pp. 351-384.

9. F. Trepel, in *Lymphozyt und klinische Immunologie* (H. Theml and H. Begemann, eds.), Springer-Verlag, Berlin, 1975, pp. 15-26.

10. F. Trepel, in *Blut und Blutkrankheiten,* Vol. 3 (H. Begemann, ed.), Springer-Verlag, Berlin, 1976, pp. 1-19.

11. K. E. Buckton, W. M. Court-Brown, and P. G. Smith, *Nature, 214,* 470 (1967).

12. G. W. Dolphin, D. C. Lloyd, and R. J. Purrott, *Health Phys., 25,* 7 (1973).

13. H. J. Evans, in *Chemical Mutagens. Principles and Methods for Their Detection,* Vol. 4 (A. Hollaender, ed.), Plenum Press, New York, 1976, pp. 1-29.

14. H. J. Evans and M. L. O'Riordan, *Mutat. Res., 31,* 135 (1975).

15. H. J. Evans, in *Handbook of Mutagenicity Test Procedures* (B. J. Kilbey, M. Legator, W. Nichols, and C. Ramel, eds.), Elsevier, Amsterdam, 1984, pp. 405-427.

16. P. Goetz, R. J. Sram, and J. Dohnalova, *Mutat. Res., 31,* 247 (1975).

17. P. Goetz, R. J. Sram, I. Kodytkkova, J. Dohnalova, O. Dostalova, and J. Bartova, *Mutat. Res., 41,* 143 (1976).

18. B. Högstedt, B. Gullberg, E. Mark-Vendel, F. Mitelman, and S. Skerfving, *Cancer Genet. Cytogenet., 3,* 185 (1981).

19. B. Högstedt, B. Gullberg, E. Mark-Vendel, F. Mitelman, and S. Skerfving, *Hereditas, 94,* 179 (1981).

20. J. A. Heddle, D. H. Blakey, A. M. V. Duncan, M. T. Goldbert, H. Newmark, M. J. Wargovich, and W. R. Bruce, in *Banbury Report 13* (B. A. Bridges, B. E. Butterworth, and I. B. Weinstein, eds.), Cold Spring Harbor Laboratory Press, New York, 1982, pp. 367-375.

21. T. Abe, T. Isemura, and Y. Kikuchi, *Mutat. Res., 130,* 113 (1984).

22. M. Littorin, B. Högstedt, B. Strömback, A. Karlsson, H. Welinden, F. Mitelman, and S. Skerfving, *Scand. J. Work. Environ. Health, 9,* 259 (1983).

23. P. Perry and S. Wolff, *Nature, 251,* 156 (1974).

24. A. V. Carrano and D. H. Moore II, in *Mutagenicity: New Horizons in Genetic Toxicology* (J. A. Heddle, ed.), Academic Press, New York, 1982, pp. 267-304.

25. P. Perry and H. J. Evans, *Nature, 258,* 121 (1975).

26. N. P. Nevsted, *Mutat. Res., 57,* 253 (1978).

27. E. Gebhart, B. Windolph, and P. Wopfner, *Hum. Genet., 56,* 157 (1980).

28. B. Lambert, U. Ringborg, E. Harper, and A. Lindblad, *Cancer Res. Rep., 62,* 1413 (1978).

29. T. Raposa, *Mutat. Res., 57,* 241 (1978).

30. P. E. Crossen, W. F. Morgan, J. J. Horan, and J. Stewart, *N.Z. Med. J., 88,* 192 (1978).

31. F. Funes-Cravioto, B. Kolmodin-Hedman, J. Lindsten, M. Nordenskföld, C. Zapata-Gayon, B. Lambert, E. Nordberg, R. Glin, and A. Swensson, *Lancet,* 322 (1977).

32. I. L. Hansteen, in *Sister Chromatid Exchange* (A. A. Sandberg, ed.), Alan R. Liss, New York, 1982, pp. 675-697.

33. A. D. Conger, *Radiat. Bot., 5,* 81 (1965).

34. M. S. Sasaki and A. Norman, *Nature, 214,* 502 (1967).

35. A. V. Carrano, *Mutat. Res., 17,* 341 (1973).

36. A. V. Carrano and J. A. Heddle, *J. Theor. Biol., 38,* 289 (1973).

37. J. Petres, D. Baron, and M. Hagedorn, *Environ. Health Perspect.* *19*, 223 (1977).

38. W. Burgdorf, K. Kurvink, and J. Cervenka, *Hum. Genet.*, *36*, 69 (1977).

39. I. Nordenson, S. Salmonsson, E. Brun, and G. Beckman, *Hum. Genet.*, *48*, 1 (1979).

40. J. Petres, K. Schmid-Ullrich, and U. Wolf, *Dtsch. Med. Wochenschr.*, *95*, 79 (1970).

41. G. Beckman, L. Beckman, and I. Nordenson, *Environ. Health Perspect.*, *19*, 145 (1977).

42. I. Nordenson, G. Beckman, L. Beckman, and S. Nordström, *Hereditas*, *88*, 47 (1978).

43. G. Beckman, L. Beckman, I. Nordenson, and S. Nordström, in *Genetic Damage in Man Caused by Environmental Agents* (K. Berg, ed.), Academic Press, New York, 1979, pp. 205-211.

44. I. Nordenson and L. Beckman, *Hereditas*, *96*, 175 (1982). '

45. A. Léonard, *Toxicol. Environ. Chem.*, *7*, 241 (1984).

46. G. Deknudt, A. Léonard, J. Arany, G. Jenar-Du Buisson, and E. Delavignette, *Toxicol. Lett.* (in press).

47. N. Degraeve, *Mutat. Res.*, *80*, 115 (1981).

48. T. H. Bui, J. Lindsten, and G. F. Nordberg, *Environ. Res.*, *9*, 187 (1975).

49. M. L. O'Riordan, E. G. Hughes, and H. J. Evans, *Mutat. Res.*, *85*, 305 (1978).

50. A. Léonard, G. B. Gerber, P. Jacquet, and R. R. Lauwerys, in *Mutagenicity, Carcinogenicity and Teratogenicity of Industrial Pollutants* (M. Kirch-Volders, ed.), Plenum Press, New York, 1984, pp. 59-126.

51. G. Deknudt, A. Léonard, and B. Ivanov, *Environ. Physiol. Biochem.*, *3*, 132 (1973).

52. G. Deknudt and A. Léonard, *Environ. Physiol. Biochem.*, *5*, 319 (1975).

53. E. Schmid, M. Bauchinger, S. Pietruck, and G. Hall, *Mutat. Res.*, *16*, 401 (1972).

54. M. Bauchinger, E. Schmid, H. Einbrodt, and J. Dresp, *Mutat. Res.*, *40*, 57 (1976).

55. Y. Shiraishi and T. H. Yoshida, *Proc. Jpn. Acad.*, *48*, 248 (1972).

56. Y. Shiraishi, *Humangenetik*, *27*, 31 (1975).

57. A. B. Bigaliev, M. N. Turebaev, R. K. Bigalieva, and M. S. Elemesova, *Genetika*, *13*, 545 (1977).

58. A. A. Azhajev, XIV Annual Meeting of the European Environmental Mutagen Society, 11-14 September 1984, Moscow, Book of Abstracts, 41 (1984).

59. K. Husgafvel-Pursiainen, P. L. Kalliomäki, and M. A. Sorsa, *J. Occup. Med., 24,* 762 (1982).

60. G. Schwanitz, G. Lehrent, and E. Gebhart, *Dtsch. Med. Wochenschr., 95,* 1636 (1970).

61. G. Schwanitz, E. Gebhart, H. D. Rott, J. K. Scholler, G. Essing, O. Lauer, and H. Prestele, *Dtsch. Med. Wochenschr., 100,* 1007 (1975).

62. I. Nordenson, G. Beckman, L. Beckman, and S. Nordström, *Hereditas, 88,* 263 (1978).

63. G. Deknudt, Y. Manuel, and G. B. Gerber, *J. Toxicol. Environ. Health, 3,* 885 (1977).

64. A. Calugar and G. Sandulescu, *Rev. Med. Chir. (Bucarest), 81,* 87 (1977).

65. A. Forni and G. C. Secchi, *Proceedings of the International Symposium on Environmental Health Aspects of Lead* (D. Barth, A. Berlin, R. Engel, P. Recht, and J. Smeets, eds.), Commission of the European Communities, Amsterdam, 1972, pp. 473-485.

66. A. Forni, G. Cambiaghi, and G. C. Secchi, *Arch. Environ. Health, 311,* 73 (1976).

67. A. Forni, *Rev. Environ. Health, 3,* 113 (1980).

68. A. Forni, A. Sciome, P. A. Bertazzi, and L. Alessio, *Arch. Environ. Health, 35,* 139 (1980).

69. A. Forni, E. Ortisi, P. Giulotto, L. Auriti, and V. Carnelli, Atti. 40° Congr. Naz. Soc. Ital. Pediatra, Palermo, October 26-30, 1979, p. 135.

70. L. D. Kathosova and G. I. Pavlenko, XIV Annual Meeting of the European Environmental Mutagen Society, 11-14 September 1984, Moscow, Book of Abstracts, *231* (1984).

71. B. Högstedt, A. M. Kolning, F. Mitelman, and A. Schütz, *Lancet,* 272 (1979).

72. K. Sperling, G. Weiss, M. Muenzer, and G. Obe, *Chromosom-Untersuchungen an Blei-exponierten Arbeitern, "Arbeitgruppe Blei" der Kommission für Umweltgefahren des Bundesgesundheitsamtes,* Berlin.

73. I. Nordenson, S. Nordström, A. Sweins, and L. Beckman, *Hereditas, 96,* 265 (1982).

74. L. Beckman, I. Nordenson, and S. Nordström, *Hereditas, 96,* 261 (1982).

75. M. L. O'Riordan and H. J. Evans, *Nature, 274,* 50 (1974).

76. J. Mäki-Paakkanen, M. Sorsa, and H. Vainio, in 29 Nordiske Yrkeshygieniske Mote i Norge, November, 3-5, 1980, Oslo, Yrkeshygienisk Institutt, 1980, p. 30.

77. E. Schmid, M. Bauchinger, S. Pietruck, and G. Hall, Mutat. Res., 16, 401 (1972).

78. M. Hoffmann, S. Hagberg, A. Karlsson, R. Nilsson, J. Ransten, and B. Högstedt (in press).

79. M. Bauchinger, J. Dresp, E. Schmid, N. Englert, and C. Krause, Mutat. Res., 56, 75 (1977).

80. M. Bauchinger, E. Schmid, and D. Schmidt, Mutat. Res., 16, 407 (1972).

81. D. Schmidt, B. Sansoni, W. Kracke, F. Dietl, M. Bauchinger, and W. Stich, Münch. Med. Wochenschr., 41, 1761 (1972).

82. P. Jacquet, A. Léonard, and G. B. Gerber, J. Toxicol. Environ. Health, 2, 619 (1977).

83. J. A. Heddle and W. R. Bruce, in Origins of Human Cancer, Vol. 4 (H. H. Hiatt, J. D. Watson, and J. A. Winsten, eds.), Cold Spring Harbor Laboratory Press, New York, 1977, pp. 1549-1557.

84. A. Léonard, G. Linden, and G. B. Gerber, in Proceedings of the International Symposium on Environmental Health Aspects of Lead (D. Barth, A. Berlin, R. Engel, P. Recht and J. Smeets, eds.), Commission of the European Communities, Amsterdam, 1972, pp. 303-309.

85. G. Deknudt and G. B. Gerber, Mutat. Res., 68, 163 (1979).

86. G. Deknudt, G. Colle, and G. B. Gerber, Mutat. Res., 454, 77 (1977).

87. P. Jacquet and P. Tachon, Toxicology, 8, 165 (1981).

88. L. A. Muro and R. A. Goyer, Arch. Pathol., 87, 660 (1969).

89. G. P. Biscaldi, G. Robustelli della Cuna, and G. Pollini, Lav. Um., 21, 385 (1969).

90. J. B. Bijlsma and H. F. De France, Int. Arch. Occup. Environ. Health, 38, 145 (1976).

91. B. Beek and G. Obe, Humangenetik, 29, 127 (1975).

92. G. R. Douglas, R. D. L. Bell, C. E. Grant, J. M. Wytsma, and K. C. Bora, Mutat. Res., 77, 157 (1980).

93. L. Dalpra, M. G. Tibiletti, G. Nocera, P. Giulotto, L. Auriti, V. Carnelli, and G. Simoni, Mutat. Res., 120, 249 (1983).

94. Q. H. Qazi, C. Madahar, and A. M. Yuceoghu, Hum. Genet., 53, 201 (1980).

95. W. Bruce and J. Heddle, Can. J. Genet. Cytol., 21, 319 (1979).

96. K. Poma, M. Kirsch-Volders, and C. Susanne, J. Appl. Toxicol., 1, 314 (1981).

97.   S. Skerfving, *Toxicology, 22,* 3 (1974).

98.   S. Skerfving, K. Hansson, and J. Lindsten, *Arch. Environ. Health, 21,* 133 (1970).

99.   S. Skerfving, K. Hansson, C. Mangs, J. Lindsten, and N. Ryman, *Environ. Res., 7,* 83 (1974).

100. R. Kato, A. Nakamura, and T. Sawai, *Jpn. J. Hum. Genet., 20,* 256 (1976).

101. H. I. Popescu, L. Negru, and I. Lancranjan, *Arch. Environ. Health, 34,* 461 (1979).

102. L. Verschaeve, M. Kirsch-Volders, C. Susanne, C. Groettenbriel, R. Haustermans, A. Lecomte, and D. Roossels, *Environ. Res., 12,* 306 (1976).

103. L. Verschaeve and C. Susanne, in *Eighth Annual Meeting of the European Environmental Mutation Society,* Dublin, July 1-13, Book of Abstracts, 1978, p. 93.

104. L. Verschaeve, M. Kirsch-Volders, L. Hens, and C. Susanne, *Mutat. Res., 75,* 335 (1978).

105. L. Verschaeve, J. P. Tassignon, M. Lefèvre, P. De Stoop, and C. Susanne, *Environ. Mutat., 1,* 259 (1979).

106. V. Mabille, H. Roels, P. Jacquet, A. Léonard, and R. Lauwerys, *Int. Arch. Occup. Environ. Health, 53,* 257 (1984).

107. M. Boysen, H. Waksvik, L. A. Solberg, A. Reith, and A. Hogetveit, in *Nickel Toxicology* (S. S. Brown and F. W. Sunderman, Jr., eds.), Academic Press, New York, 1980, pp. 35-38.

108. H. Waksvik and M. Boysen, *Mutat. Res., 103,* 195 (1982).

109. H. Waksvik, M. Boysen, and A. Hogetveit, *Carcinogenesis 5,* 1525 (1984).

110. A. Léonard and P. Jacquet, in *Nickel in the Human Environment* (F. W. Sunderman, Jr., ed.), IARC Scientific Publications No. 53, Lyon, 1984, pp. 277-291

111. M. A. Pilinskaya, *Genetika, 6,* 157 (1960).

112. N. P. Kuleshov and K. S. Sedova, XIV Annual Meeting of the European Environmental Mutagen Society, 11-14 September 1984, Book of Abstracts, *267* (1984).

113. L. V. Chapikashvili, L. A. Bolyliova, V. V. Otradnova, and O. P. Chernikova, XIV Annual Meeting of the European Environmental Mutagen Society, 11-14 September 1984, Book of Abstracts, *101* (1984).

114. K. B. Jacobson and J. E. Turner, *Toxicology, 16,* 1 (1980).

115. M. A. Sirover and L. A. Loeb, *Science, 194,* 1434 (1976).

116. J. L. Webb, *Enzyme and Metabolic Inhibitors,* Academic Press, New York, 1966.

117. J. Petres and A. Berger, *Arch. Dermatol. Forsch., 242,* 343 (1972).

118. M. Hollstein, J. McCann, F. A. Angelosanto, and W. W. Nichols, *Mutat. Res., 65,* 133 (1979).

119. A. Léonard, M. Duverger-Van Bogaert, A. Bernard, and R. Lauwerys, *Environ. Monitor. Assess.* (in press).

9

Metal Ion Carcinogenesis: Mechanistic Aspects

Max Costa[*]
Department of Pharmacology
University of Texas Medical School
Houston, Texas 77225

and

J. Daniel Heck
Lorillard Research Center
Greensboro, North Carolina 27420

---

[*]Present affiliation: Department of Environmental Medicine, New
York University Medical Center, New York City, New York 10016

1. INTRODUCTION

While it is beyond the scope of this review to discuss the data
implicating certain metals and their compounds as carcinogens,
substantial evidence exists to show that chromium, nickel, arsenic,
and perhaps also cadmium and beryllium are carcinogenic to humans
and experimental animals [1]. There are for several of these metals
examples of compounds that display either very potent or very weak
carcinogenic activity. This difference in activity appears to be
due to the bioavailability of the metal compound [2]; the more potent
compounds contain carcinogenic metal ions that enter cells readily
and are able to avidly interact with the DNA. Salts of hexavalent
chromate (CrO_4^{2-}), for example, are more potent carcinogens than are
salts of trivalent chromium $(CrCl_3)$, because the former is actively
transported into cells while the latter enters cells to only a limited extent [3]. Thus the ability of the metal to enter the cell
and the capacity of the metal ion to reach and interact with the DNA
are important molecular features of the metal carcinogenesis process
[3]. The way metal compounds enter cells may also be as important
as the quantity of metal inside the cell [4]. Evidence indicating
the importance of the mechanism of metal entry and distribution will
be presented in this chapter. A large fraction of this article involves a discussion of nickel compounds, since these agents have
been utilized extensively for mechanistic studies.

What type of cells in vivo are targets for metal carcinogenesis?
Again this depends upon the ability of the metal ion to distribute to
a cellular site and to enter the cells. Unlike organic chemicals,
the initiation of neoplasia by carcinogenic metal compounds is not
thought to require metabolic activation beyond oxidation-reduction
reactions or particle dissolution [2]. It is reasonable to assume
that if the metal can be delivered to a particular organ and appro-
priately enter the cells comprising that organ so as to produce a
sufficiently high intracellular concentration, then it can induce a
carcinogenic response. An important factor in this response involves
the overall cytotoxic activity of the metal ion. Obviously if the

metal ion is as reactive and cytotoxic as Hg^{2+}, then cell death will precede any potential carcinogenic response. An example of the relationship between metal delivery to an organ and carcinogenic response is found with lead. Although there is substantial controversy as to whether lead is carcinogenic to humans, lead acetate at reasonably high doses produces kidney tumors when fed to rats [5]. Kidney tumors generally only develop at very toxic doses of lead and generally only in animals that experience obvious acute lead toxicity (e.g., impairment of heme synthesis, kidney injury). At these doses of lead, which result in a high level of organ-specific metal accumulation in the kidney, characteristic nuclear inclusion bodies are observed in kidney cells. The kidney specificity of lead accumulation, nuclear inclusion body development, and kidney tumor induction suggests a common mechanistic relationship among these sequelae of high doses of lead [6]. Whatever its mechanism, the carcinogenic event is triggered by the accumulation of the metal in that organ consequent to the ability of the metal to enter the cell. Distribution and entry of all oncogenic metal ions into cells are key events in understanding their carcinogenic mechanisms.

2. CELLULAR UPTAKE AND INTRACELLULAR DISTRIBUTION OF METAL IONS

For the purposes of characterizing their interaction with biological systems, metals and their compounds can be divided into three categories: (a) particulate metal compounds, (b) water-soluble metal compounds, and (c) lipid-soluble compounds. The uptake of metals into cells is perhaps one of the most important determinants of carcinogenic potency among compounds of metal ions that have the capacity to induce oncogenic changes. The mechanism by which metals enter cells appears to be very important in understanding their carcinogenic activity. The details of the mechanism of entry of metal compounds into cells have not been extensively studied. Certainly a particulate metal compound that is phagocytized would enter cells by a mechanism so vastly different from that of a soluble metal ion

that dramatic effects on mechanisms of cytotoxicity and carcinogenesis may be anticipated. Whether a soluble metal ion such as
chromate (CrO_4^{2-}), which enters cells readily utilizing the SO_4^{2-}
transport system, differs in its effect on cells by virtue of its
transport mechanism alone is not known. However, because of this
uptake mechanism, intracellular levels of chromate become elevated
following exposure to this metal. Ionic nickel does not enter cells
readily and for this reason many water-soluble nickel salts are not
considered potent carcinogens. Lipid-soluble metal compounds such
as nickel carbonyl enter cells readily and for this reason are
usually very toxic. Concern over the toxicity of metal compounds
such as nickel carbonyl generally overshadows their potential carcinogenic hazards, although chronic exposure to subtoxic doses may
produce oncogenesis in the absence of overt toxicity [1].

2.1. Selective Phagocytosis of Particulate Metal Compounds

Carcinogenic metal ions derived from particulate metal compounds may
enter cancer target cells following the slow generation of water-
soluble metal species by extracellular particle dissolution or,
alternatively, by phagocytic cellular uptake of the particle itself.
The propensity of various water-insoluble particulate metal compounds
to be phagocytized by cells is of paramount importance, not only in
determining their carcinogenic activity but also in understanding
their toxicity. For example, certain metal compounds such as amorphous NiS particles display very weak carcinogenic activity in experimental animals, while other metal compounds such as crystalline Ni_3S_2
and crystalline NiS are very potent carcinogens [7]. While it has
been proposed that differences in particle solubility among inorganic
nickel compounds may contribute to their widely varying carcinogenic
potency, the dissolution rates in serum of perhaps the most potently
carcinogenic (Ni_3S_2) and least carcinogenic (amorphous NiS) of the
well-studied nickel compounds are remarkably similar [8].

An alternative hypothesis, based upon observations made in
tissue culture, proposed that the carcinogenic activity of the
crystalline nickel sulfide compounds was related to their facile
uptake by potential cancer target cells [9]. In contrast, amorphous
NiS particles were not phagocytized as readily as were the crystalline NiS particles by cells in culture, and this was associated with
a lower incidence of transformation among these cells. The active
phagocytosis of crystalline nickel sulfide compounds and the lack of
uptake of amorphous NiS particles were observed in a number of different types of cultured cells, including primary cultures of rat Sertoli
cells or human, rat, or hamster fibroblasts in primary cultures of
kidney cells and in several established cell lines such as BHK-21
cells, Chinese hamster ovary cells, and WI-38 human lung diploid
fibroblasts [10]. Studies of phagocytosis of nickel compounds have
also been conducted with mouse peritoneal macrophages by our laboratory (see discussion in Ref. 11).

The most notable difference between macrophages and facultative
phagocytes is the ability of the former to eventually take up amorphous NiS particles [11]. The rate of uptake of amorphous NiS particles is less than that of the crystalline form for both macrophages
and facultative phagocytes, but, in contrast to the macrophages, these
cells do not engulf significantly more amorphous NiS particles despite
prolonged exposure to high concentrations of these particles [2].
There is an excellent correlation among the phagocytosis of a number
of particulate nickel compounds by facultative phagocytes and their
ability to induce transformation in Syrian hamster embryo cells [11].

Hansen and Stern [12] concluded in 1983 that the observed differences in the potencies of relatively water-insoluble nickel compounds to transform BHK-21 cells were related to the bioavailability
of nickel. Since these investigators exposed their cultures to the
test particles for a brief time interval (24 hr), the dissolved nickel
generated extracellularly by even the most soluble of these particles
(i.e., crystalline Ni_3S_2) was not sufficient to induce transformation,
provided that a sufficient quantity of particles was not retained in

the culture plate following the specified exposure time interval. However, following nickel particle exposure, cells were extensively washed, trypsinized, converted to single-cell suspensions, and re-plated in both our studies [9] and those of Hansen and Stern [12]. These procedures greatly reduce the particles carried over from the exposure plates to the plates where the cells are allowed to prolif-erate and express their transformed phenotype for 2-3 weeks.

The relationship between phagocytosis and carcinogenic potency in vivo is far more difficult to examine than in tissue culture. Monolayer cultures of cells are an optimal system for observing phagocytosis of nickel sulfide particles because of the magnified physical contact of particles added over a flat sheet of cells. Nevertheless, recent studies by McCully et al. [13] demonstrated that intrarenal injection of crystalline Ni_3S_2 particles results in the phagocytosis of these particles by mesangial cells in the glomer-ulus within 1 week after particle injection. This microscopic event is followed by hyperplasia of the mesangial cells [13]. Crystalline Ni_3S_2 particles were also detected in phagocytic cells throughout the body, but their presence in potential cancer target cells and the subsequent hyperplasia of these mesangial cells suggests that this early event may be important in the subsequent abnormal division induced by Ni_3S_2 in vivo.

Support for a phagocytic mechanism of in vivo induction of tumors by Ni_3S_2 and the lack of tumor induction by similar adminis-tration of amorphous NiS may be inherent in the similar dissolution rates of these two nickel compounds. The similar dissolution rates of amorphous NiS and crystalline Ni_3S_2 in serum or other aqueous systems does not necessarily imply that their rates of dissolution within a tissue in vivo are also similar. In fact, Ni_3S_2 has been shown to undergo chemical changes when injected into tissues with the subsequent formation of crystalline NiS [14]. This latter com-pound was found to have a slower dissolution rate in serum than has amorphous NiS [8] (Table 1). A recent report indicates that 10 weeks after the intrarenal injection of Ni_3S_2 these particles were no longer

TABLE 1

Half-Times of Dissolution of Selected
Nickel Compounds[a]

Compound	H_2O	Serum
Crystalline Ni_3S_2	10.3 years	34 days
Crystalline NiS	3.3 years	2.6 years
Amorphous NiS	34 days	24 days

[a]Compounds were suspended at a concentration of 2 mg Ni/ml of H_2O or
rat serum. The particles were agitated at 37°C in either H_2O or rat
serum and the amount of dissolved nickel determined at various time
intervals.
Source: Data compiled from Ref. 8.

detectable in the animal, even at the site of injection [8]. The
concentration of [63]Ni in urine and feces following intramuscular
administration of [63]Ni_3S_2 was utilized to estimate the tissue dissolution rate of Ni_3S_2 [8]. Under these conditions the dissolution
half-time of crystalline Ni_3S_2 was reported to be 575 hr, or essentially the same as the dissolution half-time of amorphous NiS or
crystalline Ni_3S_2 in serum. Similar studies of amorphous NiS dissolution in vivo have not been published. Differences in dissolution rates either in vivo or in serum probably do not account for
the tumorigenic or transforming potency of amorphous NiS and Ni_3S_2.
Since amorphous NiS is one of the most soluble of the nickel compounds, it will produce the highest quantity of extracellular dissolved nickel in a culture plate, yet it is unable to transform
cells efficiently. Additionally, because of the slow dissolution
rate of these compounds, the dissolved nickel generated during the
limited period of exposure in cell transformation studies is well
below the level of soluble nickel required to induce transformation.
Furthermore, the incidence of transformation or carcinogenesis with
soluble nickel salts is impressively low under almost any exposure
situation.

A recent study [15] has demonstrated that repeated intramuscular injections of $NiSO_4$ (15 doses designed to model extracellular
Ni_3S_2 dissolution) did not result in tumor formation, indicating
that the simple continuous presence of extracellular ionic nickel
is not sufficient to induce carcinogenesis. The reason for this may
be because soluble ionic nickel does not enter cells readily. Several
reports have suggested that crystalline Ni_3S_2 particles administered
intramuscularly could be phagocytized by nonmacrophage cells in vivo
[14,16]. Crystalline Ni_3S_2 particles were found to be primarily
intracellular at the site of injection located in what the authors
characterized as fibroblasts and macrophages within 1 week after
injection [14,16]. Thus in vivo there is evidence of phagocytosis
of potently carcinogenic crystalline Ni_3S_2 particles. The relationship between phagocytosis and carcinogenic activity of a larger number
of nickel compounds has not been examined in facultative phagocytes.
There is good correlation among the phagocytic activities of crystalline Ni_3S_2, crystalline NiS, and amorphous NiS and of crystalline
Ni_3Se_2 and metallic Ni and the ability of these compounds to induce
transformation in Syrian hamster embryo cells and to induce carcinogenesis following intramuscular injection [10,11].

2.2. Uptake of Metal Ions and Importance
of the Metal Delivery Mechanism

A considerable amount of work has been conducted on the uptake of
nickel compounds and therefore they will be discussed in detail.
Recent studies have addressed the importance of the metal uptake
mechanism in inducing transformation with nickel compounds. The
potently carcinogenic crystalline NiS particles are phagocytized by
facultative phagocytes. The phagocytized particles are contained in
cytoplasmic vacuoles. Video intensification microscopy studies indicated that the phagocytized particles move about in the cytoplasm of
cells by saltatory motion and eventually aggregate around the nucleus
and become relatively stationary at this site [17]. Intracellular

particles dissolve at an accelerated rate compared to extracellular particles [18]. This is believed to be due to the fact that the acidic pH of the vacuoles containing the phagocytized particle results in greater particle dissolution compared to the physiological pH present extracellulary [18,19].

The cytoplasmic dissolution of phagocytized crystalline NiS particles is important for the nuclear uptake of nickel, since the metal cannot enter the nucleus in a particulate state. Phagocytized particles eventually aggregate around the nuclear membrane and release ionic nickel at this site. The fact that nickel derived from a dissolved particle can enter the nucleus has been well documented [18]. This nickel uptake mechanism differs remarkably from the uptake of ionic nickel. The mechanism of uptake of ionic nickel has not been extensively studied, but it is known that the uptake is greatly influenced by the composition of the extracellular medium [19]. The presence of metal-binding amino acids such as cysteine and histidine greatly reduces the ability of nickel as well as other metal ions to enter cells [19]. The uptake of a number of toxic or carcinogenic divalent metal ions such as Ni^{2+} or Hg^{2+} is 10-fold higher if cells are incubated in a minimal salts-glucose medium compared to McCoy's medium [19], and the uptake of these metal ions in α-minimal essential medium (α-MEM) is threefold lower than the uptake in McCoy's medium. This lower uptake in the α-minimal essential medium compared to the McCoy's medium is due to the higher concentration of metal-chelating amino acids present in the former [20]. These comparisons of metal uptake were all performed in the presence of 10% fetal bovine serum [20].

Recent studies have underlined the importance of metal uptake mechanisms in the production of specific lesions in cells [4]. One approach employed in such studies was the intracellular delivery of ionic nickel in a similar fashion as it would be delivered by phagocytosis and subsequent intracellular dissolution of a metal particle [4]. For these studies $NiCl_2$ was reacted with albumin, and any unbound ionic nickel was removed by gel filtration chromatography.

The nickel-albumin complex was encapsulated into liposomes and cultured cells were treated with this mixture [4]. Nickel delivered
into cells in this fashion was compared with nickel treatment as the
soluble metal salt, or the liposome alone, Ni complexed to albumin
alone, or crystalline NiS particles. Two endpoints were assessed
with these experiments: (a) chromosomal aberrations [4] and (b) cell
transformation. Chinese hamster ovary cells treated with water-
soluble $NiCl_2$ had several types of chromosomal aberrations, including
breaks, gaps, and exchanges [4]. There was a higher incidence of
chromosomal aberrations observed in the centromeric area of these
chromosomes, a region comprised largely of heterochromatin. Treatment of cells with crystalline NiS particles resulted in the induction of chromosomal aberrations that were similar in type and frequency to those observed with water-soluble $NiCl_2$, but in addition
to these aberrations crystalline NiS particles also caused a preferential fragmentation of the long arm of the X chromosome. In these
cells the long arm of the X chromosome is known to be heterochromatic,
and following NiS particle treatment the heterochromatic region was
preferentially damaged in the absence of any fragmentation produced
in the other chromosomes [4]. This effect was only observed when
cells were treated with crystalline NiS particles and was never found
following treatment of cells with $NiCl_2$. However, when cells were
treated with $NiCl_2$-albumin complexes encapsulated in liposomes, a
pattern of chromosomal fragmentation similar to that observed following crystalline NiS particle treatment was observed. The preferential
fragmentation of the long arm of the X chromosome appeared to be due
to the fact that during interphase heterochromatin forms the inside
lining of the nucleus [21] and is therefore the first type of DNA
encountered by nickel ions as they enter the nucleus. However, ionic
nickel is not readily taken up into cells and the limited amount of
nickel that was taken up into cells is relatively unrestricted in
terms of its interaction with various cellular ligands. In contrast,
when nickel sulfide particles are phagocytized or when nickel-albumin
conjugates are delivered into cells inside liposomes, the nickel ions

enter cells in a more restricted fashion and may in fact be directed
toward a preferential interaction with the nucleus and heterochromatin
[16,21].

The major lesion caused by nickel in the DNA appears to be the
induction of a DNA-protein crosslink [22]. This lesion, which will
be discussed in more detail in a later section of this chapter, is
relatively persistent owing to the fact that it is not readily repaired. Consequently it is likely that once it is formed, the DNA
sequences that may be blocked by this protein may not be replicated.
It is known that in intact cells a major region in the genome where
a DNA-protein crosslink will form is in heterochromatic DNA [23].
Consequently the observed chromosomal fragmentation of the heterochromatic long arm of the X chromosome may be due to the replication
of DNA over a nickel-induced DNA-protein crosslink in heterochromatin.
A less-pronounced area of DNA-protein crosslink induction is in the
euchromatin region; however, the induction of a DNA-protein crosslink
in this region may have a more acute effect on cell function, since
active genes are located and controlled in this region. The significance of the preferential fragmentation of the long arm of the X
chromosome in terms of nickel carcinogenesis is not known [4]. However, a comparison of the incidence of transformation following treatment with $NiCl_2$, crystalline NiS, and $NiCl_2$-albumin encapsulated in
liposomes indicates that with the latter the transformation frequency
is much higher than with $NiCl_2$ alone [24]. The incidence of transformation also appeared higher with the liposome-delivered nickel
compared to the $NiCl_2$-treated cells for an equivalent cellular level
of nickel. These results are only preliminary and more studies are
in order to determine their significance.

2.3. Nuclear and Nucleolar Localization
      of Carcinogenic Metal Ions

The ability of carcinogenic metal ions to enter cells and cross the
nuclear membrane is extremely important in understanding their

carcinogenicity. For example, the delivery of carcinogenic nickel into the nucleus from a phagocytized particle or by liposomal phagocytic delivery of metal ions apparently results in a high incidence of transformation. Consequently, the concentration of carcinogenic metal ions in the nucleus relative to that in the cytoplasm may be related to carcinogenesis. A high level of metal ions in the cytoplasm may cause cytotoxicity, whereas a high level of metal ions in the nucleus will cause DNA lesions and mutations, events related to oncogenesis.

Little work has been conducted on the localization of carcinogenic metals in the nucleus. The reason for this may involve problems of metal redistribution during the biochemical fractionation or fixation of tissue. Only very tightly bound metals can be studied with respect to their subcellular localization. Carcinogenic metals such as nickel and chromium are thought to form very stable ternary complexes consisting of DNA, metal, and protein [22,23,25-27]. Once formed, these complexes are extremely stable and have been implicated in the carcinogenic action of metals such as nickel and chromium [22, 23,25-27]. The redistribution of metal ions is less likely once this complex forms.

The nucleolus has been shown to preferentially accumulate a number of metal ions administered to rats in their drinking water [28], including the carcinogenic metal ions nickel and chromium. On a per-milligram-protein basis the nucleolus had 18 times more nickel than was present in the nucleus and 11.3 times the amount of chromium present in the nucleus [28]. On a similar protein-normalized basis the nucleus contained 90 times the nickel concentration present in the cell and 68 times the chromium concentration [28]. These results clearly illustrate that the nucleolus preferentially accumulates metal ions and suggests that this preferential accumulation may play an important role in nickel and chromium carcinogenesis.

3. DNA LESIONS PRODUCED BY
CARCINOGENIC METALS

Carcinogenic metals such as nickel and chromium are known to interact with DNA by binding to the phosphate groups [29]. They are known to induce three types of DNA lesions based upon analysis with the alkaline elution technique, including single-strand breaks, DNA-protein crosslinks, and DNA-DNA crosslinks [22-27]. Single-strand breaks induced by carcinogenic metal ions are known to be rapidly repaired, whereas DNA-protein crosslinks are not readily repaired and represent relatively persistent lesions [25-27,30]. It is assumed that such relatively persistent lesions are more likely to produce permanent genetic changes than lesions that are repaired rapidly. This is because the persistent lesion will be present during periods of DNA replication and this may result in alterations of DNA in daughter cells. The DNA molecule inside a single cell is 3 m long, but only about 1% of the DNA is actually expressed. Thus, if lesions such as DNA-protein crosslinks occur randomly throughout the DNA molecule, the probability that they will be induced in genetically expressed material is low. However, there is evidence that DNA lesions may not occur randomly throughout the DNA. These types of observations are quite reasonable, since the DNA molecule is not homogeneous in structure.

Recent evidence indicates that nickel induces DNA-protein crosslinks in heterochromatic regions of DNA [4,23]. Heterochromatic regions of DNA include areas where few genes are localized and expressed. The reason for this preferential crosslinking by nickel may be because of the protein found associated with the DNA present in heterochromatic regions. Nickel appeared to induce the crosslinking of proteins that have the capacity to bind tightly to DNA [25]. What types of effects may be expected as a result of the formation of DNA-protein crosslinks by carcinogenic metal ions? As indicated above, the DNA-protein crosslink is a relatively persistent

lesion and may cause problems during DNA replication. This may
result in portions of the DNA that do not replicate. The consequences of this or other effects of DNA-protein crosslinks are not
known; however, these types of lesions may be responsible for chromosomal aberrations induced by metal compounds containing carcinogenic
nickel and chromium [4]. Nickel as well as other carcinogenic metals
induced gaps, breaks, and exchanges in chromosomes [4].

The selective effects of phagocytized crystalline nickel sulfide
particles and of nickel chloride delivered in liposomes on heterochromatin, resulting in the appearance of fragmented heterochromatic
long arms of the X chromosomes in Chinese hamster ovary cells, is
also probably related to DNA lesions such as DNA-protein crosslinks
induced at that site [4]. Although the DNA-protein crosslink is a
persistent lesion that is not rapidly repaired, its formation in
heterochromatin regions ensures that it will be less rapidly repaired
than if it were formed in euchromatic regions [4]. Additionally,
carcinogenic metal ions are known to decrease the fidelity of DNA
replication, and since repair of DNA will involve DNA replication,
the presence of metal ions in the crosslink may induce mutations in
the primary sequence of DNA during repair [31]. Recent investigations on the distribution of Ni^{2+} in bacteria [32] and dinoflagellates [33-35] indicate a very high preference for its association
with genetically inactive DNA. In the dinoflagellates other potentially carcinogenic divalent metal ions preferentially associate
with genetically inactive DNA. The ability of these divalent metal
ions to preferentially and tightly bind to inactive regions of bacterial and dinoflagellate DNA is probably related to the structure
of the DNA molecule in these regions. The existence of compact DNA
conformation offering more reactive sites for divalent metal binding
as well as the protein associated with these regions may be the reasons for this tight interaction. The binding of these divalent metal
ions to these genetically inactive regions represents the major irreversible site of metal ion binding on bacteria and dinoflagellates.

4. EFFECT OF METAL IONS ON CELL GROWTH,
DNA REPLICATION, AND DNA REPAIR

The ability of chemicals to affect mitotic division and DNA replication represents potential opportunities for altering genetic composition. Such alterations of the genetic composition of cells can
lead to oncogenesis. Recent experiments have strongly suggested
that changes in DNA are involved in events related to carcinogenesis
[36]. These studies have demonstrated that transfection of human
tumor DNA into NIH 3T3 cells results in neoplastic transformation of
the cells [36]. The ability of human bladder carcinoma DNA to transform NIH 3T3 cells appeared to depend upon activation of the endogenous cellular *ras* oncogene [36]. The *ras* oncogene is altered by a
single base change which presumably results in a single amino acid
substitution in the product [36]. Recent evidence indicates that
this single amino acid change alters the function of the normal *ras*
gene product [37]. In addition to having the capacity to bind to
guanosine triphosphate (GTP), the normal *ras* product also has associated GTPase activity which hydrolyzes GTP [37]. The activated *ras*
product loses its GTPase activity. The significance of this loss in
activity in terms of neoplastic response is not known. In this particular instance a point-mutational change in DNA apparently triggered the activation of the *ras* oncogene to a gene capable of transforming normal NIH 3T3 cells [36].

In addition to direct changes in the DNA sequence resulting in
permanent mutations, metal ions may also affect mitotic segregation
of chromosomes, leading to aneuploidy. A number of carcinogenic
agents are known to interact with the kinetochores and other components of the mitotic spindle apparatus, leading to the unequal distribution of chromosomes among the daughter cells [38]. The fragmentation of a chromosome such as that occurring in the long arm of
the X chromosome by crystalline nickel sulfide may certainly lead to
unequal division of genetic material in the daughter cells, since
there is no mechanism for equal separation of multiple chromosomal

fragments [4]. Nondisjunction is known to occur with agents that
block cells in mitosis by interfering with the molecular processes
of prophase, metaphase, anaphase, and telophase. A number of metal
compounds have been shown to block cell cycle progression in mitosis
[39]. Since the separation of chromosomes during mitosis is a deli-
cately timed process, interference with these events may lead to
deposition of an extra. chromosome in one of the daughter cells.
Phagocytized particulate metal compounds may interfere with chromosomal movements directly, even though they are contained in a phagocytic vacuole and therefore not free to interact with chromosomes
during karyokinesis [17]. However, recent evidence indicates that
asbestos, as well as other inert particles that do not induce mutations, may cause aneuploidy by interference with mitosis [40].

In addition to their effect on mitosis, metal compounds have
also been shown to interfere with the DNA replication phase of the
cell cycle [41,42]. Thus carcinogenic metals such as nickel and
chromium, as well as toxic metals such as mercury and cadmium, are
known to produce a selective block in S phase, and many of these
metals at higher concentrations will also block cell-cycle progression of mitosis [41,42]. The mechanism by which metal ions such as
Ni^{2+}, Hg^{2+}, and Cd^{2+} block cells in S phase is not known. The ability
of some metal ions to produce lesions in the DNA, such as DNA-protein
crosslinks and DNA-DNA crosslinks, may be responsible for inhibition
of DNA replication [23]. However, it is unlikely that these lesions
will cause complete inhibition of DNA synthesis. A possibility that
has not been tested involves the ability of metal ions to bind to
phosphate groups of nucleotides or to the phosphodiester nucleotide
linkage of DNA and thereby inhibit DNA replication through this
interaction.

A very relevant and interesting mechanism by which metal ions
may alter genetic composition is by decreasing the fidelity of DNA
synthesis [31]. Many carcinogenic or mutagenic metals are known to
permit the insertion of incorrect nucleotides during DNA replication.
Many of the metal ions that produce lesions in the DNA also activate

DNA repair. The ability of metal ions to decrease the fidelity of repair replication may also be a mechanism for mutational changes involved with oncogenesis.

5. TUMOR-PROMOTING ACTIVITY OF METALS AND THE RELATIONSHIP BETWEEN MUTAGENESIS AND CARCINOGENESIS

Some potently carcinogenic metals such as several of the nickel compounds are not consistently detected as mutagens in mammalian [42] or in bacterial mutational assay systems [43]. However, bacteria cannot be used to assess the mutagenic activity of particulate metal compounds, since these compounds enter mammalian cells by phagocytosis and bacteria do not have this capacity. How could potently carcinogenic metals such as nickel and perhaps also arsenic possess little or no mutagenic activity? A possibility that has been suggested for nickel compounds is that they induce oncogenesis by a promotional mechanism [44]. In fact, water-soluble nickel compounds such as nickel sulfate are able to potentiate transformation in tissue culture and carcinogenesis in vivo of organic carcinogens such as benzopyrene [44]. Similar effects have also been shown for chromate [44]. These results may simply indicate that nickel and chromate induce carcinogenesis by different mechanisms than benzopyrene. These results may also indicate that benzopyrene could enhance the uptake and delivery of metal ions to oncogenically relevant sites in the cell [45]. Alternatively, it might indicate that nickel and chromium possess promotional activity, since they potentiate transformation when applied after the benzopyrene at extremely low concentrations. Further work is required to understand the mechanism of these interesting effects.

6. INHIBITION OF TRANSFORMATION AND CARCINOGENESIS BY DIVALENT METAL IONS

Several recent studies [46], as well as some earlier work [47], have demonstrated that nutritionally important divalent metal ions such as manganese, when applied at high doses, are able to inhibit metal carcinogenesis. The mechanism of this effect is not known, but these results suggest that carcinogenic metal ions may interfere with the normal function of nutritionally important divalent metal ions.

REFERENCES

1. M. Costa, *Metal Carcinogenesis Testing. Principles and In Vivo Methods,* Humana Press, Clifton, New Jersey, 1980.

2. M. Costa, *Cancer Bull., 36,* 247 (1984).

3. D. Y. Cupo and K. E. Wetterhahn, *Cancer Res., 45,* 1140 (1985).

4. P. Sen and M. Costa, *Cancer Res.* (in press).

5. G. J. Van Esch and R. Kroes, *Br. J. Cancer, 23,* 765 (1969).

6. R. A. Goyer, D. L. Leonard, J. F. Moore, B. Rhyme, and M. R. Krigman, *Arch. Environ. Health, 20,* 705 (1970).

7. F. W. Sunderman, Jr., and R. M. Maenza, *Res. Commun. Chem. Pathol. Pharmacol., 14,* 319 (1978).

8. K. Kuehn and F. W. Sunderman, Jr., *J. Inorg. Biochem., 17,* 29 (1982).

9. M. Costa and H. H. Mollenhauer, *Science, 209,* 5151 (1980).

10. M. Costa and J. D. Heck, *Adv. Inorg. Biochem., 6,* 285 (1985).

11. M. Costa, J. D. Heck, and S. H. Robison, *Cancer Res., 42,* 2757 (1982).

12. K. Hansen and R. M. Stern, *Environ. Health Perspect., 51,* 223 (1983).

13. K. S. McCully, L. A. Rinehimer, L. A. Gillies, S. M. Hopfer, and F. W. Sunderman, Jr., *Virchows Arch.* (in press).

14. A. Oskarsson, Y. Anderson, and H. Tjalve, *Cancer Res., 39,* 4175 (1979).

15. K. S. Kasprzak, P. Gabriel, and K. Jarczeaska, *Carcinogenesis, 4,* 275 (1983).

16. F. W. Sunderman, Jr., K. S. Kasprzak, T. L. Lau, P. P. Minghetti, R. Maenza, M. Becker, C. Onkelinx, and P. J. Goldblatt, *Cancer Res., 36,* 1790 (1976).

17. R. M. Evans, P. J. A. Davies, and M. Costa, *Cancer Res., 42,* 2727 (1982).

18. M. P. Abbracchio, J. S. Hansen, and M. Costa, *J. Toxicol. Environ. Health, 9,* 663 (1982).

19. M. P. Abbracchio, R. M. Evans, J. D. Heck, O. Cantoni, and M. Costa, *Biol. Trace Element Res., 4,* 289 (1982).

20. O. Cantoni, N. T. Christie, S. H. Robison, and M. Costa, *Chem. Biol. Interact., 49,* 209 (1984).

21. T. C. Hsu, *Genetics, 79,* 137 (1975).

22. R. B. Ciccarelli, T. H. Hampton, and K. W. Jennette, *Cancer Lett., 12,* 349 (1982).

23. S. R. Patierno, M. Sugiyama, J. P. Basilion, and M. Costa, *Cancer Res.* (in press).

24. C. Miller and M. Costa (in preparation).

25. R. B. Ciccarelli and K. E. Wetterhahn, *Cancer Res., 44,* 3892 (1984).

26. R. B. Ciccarelli and K. E. Wetterhahn, *Cancer Res., 42,* 3544 (1982).

27. J. E. Lee, R. B. Ciccarelli, and K. W. Jennette, *Biochemistry, 29,* 771 (1982).

28. H. Ono, O. Wada, and T. Ono, *J. Toxicol. Environ. Health, 8,* 947 (1981).

29. L. G. Marzilli, in *Metal Ions in Genetic Information Transfer* (G. L. Eichhorn and L. G. Marzilli, eds.), Elsevier/North-Holland, New York, 1981, p. 48.

30. M. J. Tsapakos, T. H. Hampton, and K. E. Wetterhahn, *Cancer Res., 43,* 5662 (1983).

31. M. A. Sirover and L. A. Loeb, *Science, 194,* 1436 (1976).

32. R. H. Al-Rabee and D. C. Sigee, *J. Cell Sci., 69,* 87 (1984).

33. L. P. Kearns and D. C. Sigee, *Experientia, 35,* 1332 (1979).

34. L. P. Kearns and D. C. Sigee, *J. Cell Sci., 46,* 113 (1980).

35. D. C. Sigee and L. P. Kearns, *Cytobios, 31,* 49 (1981).

36. M. McCoy, J. Toole, J. M. Cunningham, E. S. Chang, B. R. Lowy, and R. A. Weinberg, *Nature, 302,* 79 (1983).

37. R. K. Wierenga and W. G. J. Holl, *Nature, 302,* 842 (1983).

38. T. C. Hsu, *Symposium in Aneuploidy,* Washington, D.C., 1985.

39. P. K. Basrur and J. P. W. Gilman, *Cancer Res., 27,* 1168 (1967).

40. T. W. Hesterberg and J. C. Barrett, *Cancer Res., 44,* 2170 (1984).

41. M. Costa, O. Cantoni, M. deMars, and D. E. Swartzendruber, *Res. Commun. Chem. Pathol. Pharmacol., 38,* 405 (1982).

42. D. M. Skilleter, R. J. Price, and R. F. Legy, *Biochem. J., 216,* 773 (1983).

43. J. D. Heck and M. Costa, *Biol. Trace Element Res.*, *4*, 319 (1982).

44. E. Rivedal and T. Sanner, *Cancer Res.*, *41*, 2950 (1981).

45. F. W. Sunderman, Jr., *Fed. Proc.*, *37*, 40 (1978).

46. F. W. Sunderman, Jr., and K. S. McCully, *Carcinogenesis, 4*, 461 (1985).

47. F. W. Sunderman, Jr., T. J. Lau, and L. F. Cralley, *Cancer Res., 34*, 92 (1974).

10

Methods for the in vitro Assessment of Metal Ion Toxicity

J. Daniel Heck
Lorillard Research Center
Greensboro, North Carolina 27420

and

Max Costa[*]
Department of Pharmacology
University of Texas Medical School
Houston, Texas 77225

[*]Present affiliation: Department of Environmental Medicine, New York University Medical Center, New York City, New York 10016

1. INTRODUCTION

1.1. In Vitro Toxicology

Recent years have brought on an increasing concern for the potential
toxic effects of incidental or intentional exposures to the myriad
substances which constitute the products and by-products of our
advancing technologies. An acute need for the toxicological characterization of the organic, inorganic, and physical components of our
environment has arisen so that the liabilities of their utilization
can be minimized. The sheer mass of toxicological data required
both by prudence and by governmental regulatory agencies to satisfy
this need has stimulated the development of informative, rapid, and
cost-efficient in vitro methods for toxicological assessment.

The in vitro study of biological phenomena offers a number of
advantages over traditional in vivo models employing live experimental
animals. Protocols employing prokaryotic or eukaryotic cell cultures
as models of basic biological phenomena frequently offer advantages
in terms of time and cost efficiency. Studies of mechanisms or documentation of subtle toxic effects on cells may be far more practical
in in vitro model systems in which the metabolic, circadian, hormonal,
neurophysiological, and respiratory variables inherent to whole-
animal systems are eliminated or controlled. The mechanisms of action
of many toxic substances have been elucidated only because advances
in mammalian cell culture and molecular biology have permitted the
requisite degree of experimental control and manipulation. Beyond
such practical considerations, an increasing perception on the part
of the general public, as well as the scientific community, of the
desirability of avoiding unnecessary animal experimentation has provided support for the development of in vitro toxicological assessment methodologies as alternatives to traditional animal models of
toxicological phenomena.

Genetic toxicology has been at the forefront in the development
of in vitro testing methodologies, largely because traditional in vivo
bioassays to assess carcinogenesis are exceptionally costly and time-

consuming. Such chronic animal carcinogenesis bioassays carry a
cost of approximately $500,000, require several years to complete,
and not infrequently yield equivocal or controversial results which
are subject to criticism on the basis of interspecies metabolic
differences or dose-dependent pharmacokinetic phenomena [1,2]. The
urgent need to assess the genotoxic potential of the hundreds of
substances entering commerce annually has driven the development of
microbial and mammalian cell systems to supplement the cumbersome
in vivo carcinogenesis bioassay. Many of the basic techniques and
in vitro model systems developed for genetic toxicology assessment
are also applicable both to the study of other toxicological endpoints and to the task of expanding out basic knowledge of cellular
physiology, metabolism, and molecular genetics. The in vitro assessment of the toxicity of metal ions may thus serve to expand our
knowledge both of their mechanisms of action in biological systems
and of the normal function of the cell itself.

1.2. Metal Ions in In Vitro Systems

Metallic compounds have in the past received less attention as subjects for in vitro toxicological assessments than organic compounds
such as drugs, pesticides, and carcinogenic hydrocarbons. An increasing appreciation of the potential roles of metals as toxic and
carcinogenic hazards, however, is evident in the recent rapid expansion of in vitro research and testing in this area. Recent reviews
of progress in metal carcinogenesis attest to the prominent contribution of in vitro methodologies in this area [3,4]. In vitro
studies of metal-induced mutagenesis [5], neoplastic transformation
[6], and metal-macromolecule interactions [7] have been the subjects
of recent reviews. Hansen and Stern [8] compiled a comprehensive
review of all in vitro genetic toxicology studies of metal compounds
published through the end of 1982. The utility of in vitro systems
for metal ion toxicological assessment has now been firmly established.

Metal ions are in a number of respects particularly well
suited for study in vitro. Sensitive analytical methods such as
atomic absorption spectroscopy and x-ray fluorescence spectrometry
facilitate accurate quantitation of metal ions on a subcellular
scale. The availability of radionuclides of the toxicologically
significant metals makes possible metabolic, dispositional, and
mechanistic studies of extreme sensitivity. The comparatively
simple cellular biotransformation of metal compounds is generally
limited to dissolution, precipitation, macromolecular binding, and
changes in oxidation state, phenomena which likely occur in vitro
in a manner quite similar to those occurring in vivo. A primary
limitation of in vitro toxicology assessment systems has been their
inability to accurately model the complex and interrelated enzymatic
activation and detoxification pathways so important in the toxicology
of organic compounds [9]. Mammalian xenobiotic metabolism has traditionally been modeled in vitro by the addition of a subfraction of a
liver homogenate having a variety of enzyme activities [10] or, more
recently, by the utilization of primary cultures of hepatocytes
having largely intact xenobiotic metabolizing capabilities [11,12].
While adequate for toxicological screening purposes, these in vitro
biotransformation systems cannot be expected to model every element
of in vivo xenobiotic metabolism with quantitative and qualitative
fidelity. While enzymatic processes have been implicated in the
biotransformation of some metal compounds [13,14], the inadequacies
of in vitro metabolic systems currently employed in toxicological
assessments are much less significant in the testing of metals than
they are for organic compounds. The relative simplicity of the biotransformation of metal compounds ensures that meaningful and informative studies of quantitative, qualitative, and mechanistic aspects of
metal ion toxicity are practical in the variety of available in vitro
systems.

2. BIOCHEMICAL METHODS

2.1. Biochemical Assessment of Metal Ion
Cytotoxicity

An extensive battery of clinical chemistry tests is available to the
in vivo toxicologist for the assessment of toxic effects in experimental animals. Clues as to toxic mechanisms, target organ specificity, accumulation of chronic toxic effects, and the development
of tolerance may be available to the perceptive investigator who
monitors biochemical changes accompanying exposure to toxic agents
[14]. Many of these same parameters may be monitored in vitro to
provide clues to toxic mechanisms at the cellular level. Basic biochemical parameters of cytotoxicity include measurement of cell membrane integrity by vital dye exclusion [15] or by the intracellular
retention of cytoplasmic enzymes [16] or a ^{51}Cr isotopic marker [17].
Measurement of cellular adenosine 5'-triphosphate provides a sensitive
index of cell viability and metabolic status [18], while monitoring
of cellular enzymatic activities for toxicant-induced stimulation
[19], inhibition [20], or dysfunction [21] may be mechanistically
informative and may permit the exploitation of these toxicological
and pharmacological effects as research tools.

Prominent among biochemical methods for the assessment of metal
ion toxicity are those studies directed toward elucidating the role
of cellular metal-binding proteins. Metallothionein, a cysteine-rich,
low molecular weight metalloprotein thought to be involved in normal
copper and zinc homeostasis [22], also binds a number of important
toxic metals with high affinity [23]. Exposure of cells to these
metal ions (cadmium, mercury, zinc, copper) induces the synthesis of
metallothionein, providing an intracellular sink of cysteine sulfhydryl groups capable of sequestering metal ions and preventing their
binding to critical nucleophilic sites on biological macromolecules
[23,24]. A number of assays for the quantitation of this interesting
protein have been described [25-28], some particularly applicable to

sensitive metallothionein determination on the small scale often
required for in vitro studies [29-31].

The central role of metallothionein in the detoxification of
heavy metal ions and the phenomenon of metallothionein induction by
these ions have been clarified as a result of work in in vitro systems. Durnam and associates have prepared a c-DNA probe using the
pBR322 plasmid vector in *Escherichia coli* and have employed it to
isolate, characterize, and sequence a metallothionein gene whose
tissue-specific transcriptional regulation by metals was demonstrated
in an in vivo/in vitro mouse model [32-34]. The induction of metallothionein and its role in preventing Cd(II) toxicity has also been
amply demonstrated in cultured cell lines [35,36]. These in vitro
studies of the metallothionein gene and the utility of toxic metal
ions as inducers of its expression have contributed significantly to
our understanding of both metals toxicology and basic molecular
biology [32-37].

Details of the interactions of metallothioneins with toxic
metal ions have been revealed by observations of such interactions
in vitro [22,38], while an ex vivo isolated, perfused liver system
was employed by Day and associates to exploit the advantages of both
in vivo and in vitro methodologies [39]. The utilization of mammalian
cell cultures in the study of metallothionein induction and its role
in metal detoxification has proven highly informative and is discussed
below.

2.2. Biochemical Assessment of Metal Ion Genotoxicity

The effects of toxic metal ions on the structure and function of DNA
are many and varied, as might be expected for a biological macromolecule whose normal structure and function are dependent upon physiological ions and whose component nucleotides offer numerous potential
metal-binding sites. Nonspecific or specific metal binding, altered
helix configuration, helix stabilization or destabilization, and

regulation of genetic expression are among the effects of metal ions on DNA amenable to in vitro investigation [40-42].

An in vitro DNA polymerase-template system was employed to assess the effect of various metal ions on the function of the Zn(II)- and Mg(II)-requiring polymerase enzyme responsible for DNA synthesis [43,44]. A number of metals were found to be capable of activating polymerization in place of the normally required Mg(II). Several metal ions experimentally and epidemiologically associated with cancer development decreased the fidelity with which the polymerase enzyme discriminated the proper nucleotide for inclusion in the nascent DNA chain [44]. The resultant base mispairings generated by certain metal ions in this in vitro system may reflect their ability to induce DNA damage, DNA repair, mutagenesis, and carcinogenesis in other in vitro and in vivo systems. These investigations demonstrate the fine degrees of experimental control and manipulation required to identify possible mechanisms of toxicity at the molecular level. Such control is possible only in vitro, assuring a continuing need for expansion of in vitro capabilities in the future.

3. MICROBIOLOGICAL METHODS

The basis for the development of bacterial models to study toxicological phenomena relevant to humans involves an appreciation for the fact that biological systems, from bacteria to humans, utilize DNA as their genetic material and employ somewhat similar molecular machinery to synthesize, maintain, and utilize the biological macromolecules required for life. Microbial systems have thus far been applied most extensively in the assessment of genetic toxicology and in this endeavor have become essential to modern in vitro toxicological research and testing. The significant although not perfect correlation between mutagenic activity in bacteria and carcinogenic activity in experimental animal and human epidemiological studies reported for many classes of chemical compounds attests to the validity of using such rapid and inexpensive in vitro methods as

preliminary screening tests so that the limited resources available
for long-term animal studies may be used most efficiently [45].

The *Salmonella typhimurium* reverse mutation assay system
developed by Ames and colleagues [46,47] is the best-validated and
most widely applied of the bacterial mutagenesis assay systems.
The *Bacillus subtilis rec* assay, which detects DNA-damaging compounds
by their enhanced toxicity in a DNA-repair-deficient (*rec⁻*) bacterial
strain, represents another approach to the identification of potential
carcinogens [48]. A number of other toxicological test systems employing prokaryotic or eukaryotic microorganisms have been described
and their applications have been reviewed [49-51].

The application of microbial systems to the task of assessing
and studying metal ion toxicity has perhaps been less broadly successful in reflecting the results of animal studies and epidemiological
data than has similar screening of organic compounds. A number of
metals (As, Be, Cd, Ni) having suspected or confirmed carcinogenic
potential in vivo are either inactive or inconsistently detected as
active in the well-validated bacterial mutagenesis assays [5,52-55].
Only chromate has been consistently detected as a bacterial mutagen
among inorganic metal compounds strongly suspected of being human
carcinogens. New or improved bacterial testing protocols have in
some instances proven more successful in detecting metal ion genotoxicity, offering hope that specialized tests of this class may be
available in the future [56,57].

A number of plausible hypotheses to account for the inconsistent results observed for metals in bacterial mutagenesis assays have
been proposed. The precipitation of toxic metal ions by bacterial
medium components has been observed [3,52,57]. The toxic or bacterio-
static properties of metal ions may inhibit the growth and detection
of mutant colonies in these systems [58,59]. The inability of some
metal ions to efficiently penetrate the bacterial cell wall or the
possibility that metals may act as comutagens rather than as primary
mutagens has also been proposed to account for their inconsistent
detection in bacterial assays which have proven to be responsive to

many classes of organic carcinogens. Indeed, with regard to this latter hypothesis, various metals have been observed to enhance the bacterial mutagenicity of ultraviolet light [60], alkylating agents [61], azide [62], and nitrosamines [63]. The capacity of metals to inhibit efficient repair of spontaneous or carcinogen-induced DNA damage may account for such metal synergism in vitro and may contribute significantly to their in vivo genetic effects.

The interaction of metals with other mutagens does not, however, always result in a genotoxic synergism. It has been shown that $CoCl_2$ exerts an inhibitory effect on bacterial mutation induction by amino acid pyrolysates [64], N-methyl-3-nitro-1-nitrosoguanidine (MNNG) [65], and gamma radiation [66], as well as on spontaneous mutation [67]. Manganese, iron, and copper were reported to reduce the mutagenicity of quercetin in *Salmonella* strain TA98, an effect likely mediated by enhanced catalytic oxidation of quercetin [68]. Finally, selenium has been shown on a number of occasions to exert antimutagenic effects in bacterial systems [69], an observation consistent with reports of anticarcinogenic effects of this essential trace nutrient in vivo [70].

Thus, while the existing well-validated bacterial mutagenesis protocols do not in their present form appear to be applicable to effective screening of metal compounds, these or other bacterial systems have utility in the study of some basic aspects of the genotoxic activity of metals. The bacterial cell model offers the advantages of relative simplicity and access to a vast library of genetic and enzymological data which may greatly facilitate studies of metal-DNA interaction and metal-induced DNA damage and repair.

4. MAMMALIAN CELL CULTURE METHODS

4.1. Assessment of Metal Ion Cytotoxicity

Advances in mammalian cell culture technology have provided numerous systems and approaches for both routine toxicological screening and

for mechanistic studies at the cellular and molecular level.
Eukaryotic cell culture is now indispensable in toxicological
assessments of both organic and inorganic compounds. For decades
cell and molecular biologists have employed a number of easy-to-
maintain permanent cell lines as basic research tools and as a
result have provided a library of biochemical, structural, and
genetic data readily applicable to mechanistic studies of metal
ion toxicity in vitro.

Mammalian cell culture has been successfully employed in
modeling every aspect of the interaction of toxic metal ions with
cells, including metal uptake [71,72], intracellular dissolution
[73], intracellular distribution and disposition [7,73,74], macromolecular binding [74], and effects on cellular growth [75,76] and
respiration [18]. The effects of metal ions on enzymatic activities
frequently play prominent roles in their mechanisms of toxicity.
Cultured cells serve as effective systems for the study of these
effects free from the influences of dynamic hormonal, circadian,
and respiratory variables. The mechanisms of certain of the toxic
effects of metal ions, notably their genotoxic activity, are largely
unknown and are most effectively studied in intact cell systems.
The advancement of both toxicological and basic biological knowledge
has proceeded apace in the last two decades, largely owing to discoveries made possible by the evolution of experimental capabilities
in mammalian cell culture.

The initial step in metal ion toxicity, that of cell entry, is
quite amenable to in vitro study. An explanation for the potent
toxicity of oxyanions such as chromate in contrast to the much lower
toxic potential of analogous metal ions of lower oxidation states
was provided by the in vitro demonstration of the ability of chromate,
vanadate, molybdate, tungstate, and arsenate to enter cells efficiently via the sulfate and phosphate active transport systems [20].
The physical forms of various binary compounds and salts of nickel

were shown to have a profound effect on their uptake and subsequent toxicity in cell cultures. The potently carcinogenic crystalline nickel sulfides NiS and Ni_3S_2 were observed to be avidly phagocytosed in cultured cells, while the noncarcinogenic amorphous form of NiS was not as actively taken up by similar cell cultures [77,78]. Ions of Ni(II) derived from soluble nickel salts such as $NiCl_2$ were shown to enter cells far less efficiently than nickel borne in engulfed particles [71], an in vitro observation reflecting their lack of carcinogenic potential when compared to the relatively insoluble particulate nickel compounds [79]. The ability of mammalian cell culture systems to accommodate inorganic particulates and metal compounds of relatively low solubility which may efficiently deliver large amounts of toxic metal ions into cells following their phagocytosis is important, since metal particulates and fumes of respirable size constitute one of the most significant carcinogenic hazards of metals to humans. Accumulating evidence attests to the importance of the physical form and chemical composition of metal compounds as well as the oxidation state and intrinsic bonding (hard/soft) character of their component metal ions in determining both toxicity and carcinogenicity [20,72,76,80]. The dramatic effects of these variables on the capacity of metal ions to enter cells and thereby gain access to toxicologically important intracellular sites emphasizes the need for in vitro toxicological assessments employing intact cells.

A most interesting approach to the in vitro assessment of metal ion cytotoxicity in mammalian cells has been made possible by the development of cadmium-resistant cell lines by Hildebrand and associates [81]. Fine details of the mechanism of cadmium cytotoxicity and of the key role of metallothionein in the intracellular disposition and detoxification of cadmium have been revealed in these continuing studies [82,83]. The possibility that similar approaches may be employed to study the toxic mechanisms of other metal ions offers opportunities for future in vitro research.

4.2. Assessment of Metal Ion Genotoxicity

4.2.1. DNA Damage and Repair

By virtue of its role as the repository for all structural, regulatory, and developmental information in every cell, DNA constitutes a
critical target for any potentially toxic compound. Maintenance of
the integrity of DNA is essential to, and in many respects synonymous
with, the continuation of life. A number of factors contribute to
the particular vulnerability of this most critical of biomolecules
to perturbation by metal ions. The extremely large size of the DNA
polymer offers an abundance of potential metal-binding sites, many
of which are normally occupied by physiological ions essential for
proper function and for maintenance of complex higher orders of
structure [40,41]. The role of physiological ions in the functioning of enzymes participating in DNA synthesis and repair, RNA synthesis and post-transcriptional modification, and protein synthesis
provide yet more potential sites for toxic metal binding with subsequent disturbances in these essential functions. The investigation
at the cellular and molecular levels of the potential of metals to
induce DNA damage and subsequent DNA repair has been made possible
by the advancement of in vitro experimental capabilities in molecular
and cell biology. The application of these elegant in vitro methodologies to the investigation of the mechanisms of metal ion toxicity
has yet to be fully pursued and holds great promise as an avenue for
future research.

An extremely sensitive technique for the in vitro assessment
of DNA damage has been described by Kohn and associates [84]. Application of this alkaline elution technique or its variants permits the
detection of single-strand DNA breaks, double-strand breaks, inter-
strand DNA crosslinks, alkali-labile sites, DNA-protein crosslinks,
and DNA repair. The alkaline elution method exhibits many of the
best characteristics of an in vitro toxicological assessment system:
It is adaptable to comparative studies utilizing both in vivo and
in vitro exposure conditions, it may provide clues to the mechanism
of a compound's genotoxicity, results are obtained rapidly (within

48 hr of experimental treatment) and at a reasonably low cost, and
the method is quite sensitive, capable of detecting approximately
1 DNA lesion per 10^7 nucleotides [84]. While the physical details
of alkaline DNA elution are only incompletely understood, the process
is conceptually simple: Damaged (fragmented) DNA is eluted through
a filter more rapidly than undamaged DNA, while DNA crosslinked to
other DNA strands or to proteins is eluted more slowly than undamaged
(control) DNA. The rate of DNA passage through the filter pores
under specified conditions thus provides qualitative and quantitative
information about DNA damage arising from chemical exposures. The
alkaline elution technique has been applied to identify the target
organs, the time course of appearance, and the repair of nickel-
induced DNA strand breaks and crosslinks in the rat in an elegant
series of in vivo/in vitro studies [85,86]. Chromate was similarly
shown to induce DNA crosslinks in liver and kidney in vivo [87],
while Cr(VI) and Ni(II) induced both DNA-protein crosslinks and DNA
single-strand breaks in metal-treated cell cultures examined with
the alkaline elution technique [88-90]. DNA repair synthesis following in vitro exposures of cells to these carcinogenic metal ions as
well as to highly toxic Hg(II) was studied in DNA-repair-competent
Chinese hamster ovary (CHO) cells and in their repair-deficient
derivative line EM9 by following changes in DNA elution rates over
time [90]. In these studies DNA was identified as the intracellular
target relevant to the cytotoxic effects of Cr(VI) and Hg(II), a
situation exacerbated by mercury's apparent inhibition of DNA repair
enzyme activity [90]. The alkaline elution technique has also been
applied to in vitro studies of the mechanism of cadmium's genotoxicity: $CdCl_2$-induced DNA damage was characterized as single-strand
scissions subject to rejoining by an excision repair mechanism [91]
and whose generation is apparently associated with active oxygen
species [92]. A similar involvement of active oxygen species in
mercury-induced cytotoxicity has also been described [93].

Other in vitro methods for the assessment of metal-induced DNA
damage and repair have been applied successfully. Sucrose and cesium

chloride density gradient centrifugation has been employed in the
study of DNA damage [94-96] and repair [97] following treatment
of cultured cells with nickel compounds. The measurement of un-
scheduled DNA synthesis by quantitation of the incorporation of
radioactive nucleotides into the DNA of cells not in the S phase
of the cell cycle holds significant promise as a method for the in
vitro assessment of the repair of chemically induced DNA damage,
but has not as yet been extensively applied in the testing of metal
ions [98].

The assessment of metal ion-induced DNA damage and repair
appears to be a fruitful avenue for future research. The pursuit
of these investigations in vitro using mammalian cell culture will
continue to provide clues to the mechanisms by which cells maintain
their genetic integrity and by which metal ions affect DNA structure
and function.

4.2.2. Chromosomal Aberrations and Sister
Chromatid Exchanges

Chromosomal fragmentation, rearrangement, and numerical abnormalities
have been strongly associated with the development of neoplastic and
genetic diseases [99]. The capacity of chemicals to induce such
chromosomal aberrations (CAs) in cultured mammalian cells has been
demonstrated to correlate well with a known or suspected carcinogenic potential [100,101]. Sister chromatid exchanges (SCEs), the
exchange of segments of chromosome arms at apparently homologous
loci, may be detected by differential staining of sister chromatids
in metaphase cells [102]. Assessment of the induction of SCE by
chemicals has been proposed and utilized as a rapid and sensitive
in vitro assay to detect suspected carcinogens [100,103]. These
assay procedures offer the advantages of ready adaptability to in
vivo experimental animal or human biomonitoring applications. Both
CA and SCE assays have been utilized to assess metal ion genotoxicity
[8,104,105]. These methods for detecting genetic effects at the
chromosomal level exploit the advantages of the use of mammalian

target cells and share many of the most desirable characteristics of
in vitro toxicological study in terms of rapidity, low cost, sensitivity, and selectivity.

The continued application of CA and SCE analysis to the in
vitro assessment of metal ion genotoxicity may also prove informative
in elucidating the mechanisms and defining the consequences of these
expressions of genetic toxicity. Recent studies have demonstrated
that particles of potently carcinogenic crystalline NiS, which are
actively phagocytosed by cells, induce chromosome damage preferentially in heterochromatin regions [106]. This damage, manifest as
fragmentation of the X chromosome in CHO cells, was not observed
after treatment of the cells with water-soluble nickel salts such
as $NiCl_2$. Treatment of cells with a liposome-encapsulated $NiCl_2$-
albumin complex, however, induced a similar degree of chromosomal
fragmentation as was observed following treatment with the phagocytosed crystalline NiS particles. The preferential interaction
of nickel derived from phagocytosed NiS particles or liposome-
encapsulated $NiCl_2$-albumin complexes with heterochromatin may result
from the location of heterochromatin near the inner surface of the
nuclear membrane in the interphase cell [107]. Nickel derived from
particles residing in perinuclear cytoplasmic vacuoles thus en-
counters heterochromatin first upon entering the nucleus.

4.2.3. Mammalian Cell Mutagenesis

Mammalian cell mutagenesis assays offer the advantages of measuring
an endpoint highly correlated with carcinogenesis in cell lines
which more closely resemble the target cells of in vivo carcinogenesis than do the bacteria used in the widely applied prokaryotic
mutagenesis assays. Among the best validated of the mammalian cell
mutagenesis systems are those which detect forward mutations at the
HGPRT locus on the X chromosome of CHO cells [108] and at the thymi-
dine kinase (TK) locus of a mouse lymphoma cell line, L5178Y $TK^{+/-}$,
heterozygous at this locus [109]. The application of these and other

mammalian cell mutagenesis assays has been reviewed and evaluated [110-112]. The mutagenesis screening of metal compounds has also been separately reviewed [5,8].

The advantages of mammalian cell systems for in vitro toxicity assessment discussed previously are evident in the substantial correlation between eukaryotic cell mutagenesis and in vivo carcinogenesis by metal ions. The structural and functional similarities of the mammalian cell lines employed in mutagenesis assays and the target cells of in vivo carcinogenesis are substantial, and prospects for the development of epithelial cell culture technology promise even better modeling of metal ion genotoxicity in the future [113].

The mammalian cell mutagenesis assays, like other in vitro and in vivo models of toxicological hazard, are undergoing refinement and development even as they are being applied. It is therefore only natural to expect that theoretical and practical adjustments will be made as test data accumulate. An interesting example of such a development which may be relevant to in vitro metal ion mutagenesis is the recent reports that reduction in the pH of the cell culture medium may be capable under some conditions of producing mutations in the L5178Y TK$^{+/-}$ mouse lymphoma cell assay [114-116]. Acidic pH shifts or excessively high concentrations of even nontoxic physiological ions such as K$^+$ and Na$^+$ have also been shown to produce suspected false-positive results in other genetic toxicity assays employing mammalian cells [116,117]. The mechanism of this phenomenon of mutation induction under conditions of acidic pH or excessive ion concentrations is presently unknown and under active investigation. The possibility that metal ions might induce mutation or chromosomal aberration by artifactual ionic or osmotic effects in mammalian cell systems in which compounds are traditionally tested at levels up to and including those showing significant cytotoxicity should be borne in mind. Whether this phenomenon has relevance to any in vivo manifestation of metal ion toxicity is presently not known.

4.2.4. Mammalian Cell Neoplastic Transformation

A number of in vitro mammalian cell morphological transformation
assay systems have been developed as models for the study of changes
at the cellular and subcellular levels which accompany the transformation of a normal cell into a cancer cell [118]. Transformation assay
systems may employ either freshly isolated primary cultures, as in
the Syrian hamster embryo (SHE) cell system [119]; they may employ
established cell lines such as the C3H10T 1/2 [120] and BALB/c 3T3
mouse cells [121]; or they may employ cells infected with oncogenic
retroviruses as in the SHE cell/SA7 virus system [122]. Each transformation assay system offers certain advantages and all share a
common endpoint in the quantitation of phenotypic changes in cells
exposed to carcinogenic agents. The phenotypic changes which signal
morphological transformation in vitro reflect characteristics of
cancer cells in vivo and herald the initiation of a sequence of
events culminating in the acquisition of a capacity to form tumors
when inoculated into suitable host animals. The application of
transformation assay systems to the task of screening various substances for potential carcinogenicity has been recently reviewed
[123,124]. Transformation assay protocols are constantly undergoing
improvement and refinement. The establishment of assay systems
utilizing human or epithelial cells appears feasible in the future.
The phenomenon of in vitro morphological transformation of cells
following their exposure to carcinogens constitutes an endpoint
which is arguably closer to the in vivo carcinogenic event than any
other endpoint measurable in vitro. It is therefore particularly
significant that metal compounds of confirmed or suspected carcinogenic potential have been efficiently detected as positive in transformation assays [6,8,124].

The mammalian cell transformation systems are particularly
well suited to assessments of metal compounds by virtue of their
ability to accommodate even relatively insoluble particulate metal
compounds which are in many cases far more toxic and carcinogenic
than soluble metal salts. A well-characterized example of this is

found among inorganic nickel compounds. Soluble nickel salts such as $NiCl_2$ and $NiSO_4$ are essentially nontumorigenic in experimental animals, while several crystalline nickel sulfides are exceedingly potent in this regard [79,125]. Another particulate nickel compound, amorphous NiS, is lacking in both carcinogenic and transforming potency [6,77,78,96,125,126].

The selective phagocytic uptake of the crystalline nickel sulfides NiS and Ni_3S_2 has been correlated with their potent geno-toxic, carcinogenic, and transforming activities, while the low geno-toxic potential of amorphous NiS was consistent with observations of this compound's inability to trigger its uptake by phagocytosis into the cell [72,77,78,96,126,127]. The demonstration of the importance of particle phagocytosis as a mechanism for efficient delivery of genotoxic metal ions into the cell in determining carcinogenic and transforming potency among a series of nickel compounds underscores the advantages of intact cell systems in vitro.

Future prospects for in vitro transformation assay systems in the assessment of metal ion toxicity include the identification of transforming genes and gene products induced by carcinogenic metals. Metal ions may prove uniquely useful in the search for the underlying mechanism of neoplastic transformation and await full utilization toward this goal.

5. GENERAL CONCLUSIONS

Advances in the fields of cell biology, genetics, and molecular biology have been made possible by the development of in vitro methodologies to study and model their mechanistic aspects at the cellular and subcellular levels. The application of these same in vitro techniques to assess toxicological phenomena has come into prominence more recently as a result of both societal needs and the scientific utility of the methods themselves. The in vitro assess-ment of the cytotoxic and genotoxic potential of chemicals is now an integral component of modern toxicology.

While the presently available bacterial mutagenesis assay systems have not been shown to reliably reflect the genotoxic potential of metals in vivo, most other in vitro systems have proven to be effective and informative in the study of metal ion toxicity and mechanisms. Mammalian cell culture systems have proven to be particularly effective in these regards. A spectrum of metal ion-induced genotoxic effects, including altered DNA structure, DNA damage and repair, mutagenesis, and transformation, have been identified only as a result of advances in in vitro methods and techniques. Continued progress in the development of in vitro experimental capabilities ensures that a more complete understanding of the toxic effects of metal ions will be attained in the near future.

REFERENCES

1. B. L. Van Duuren and S. Banerjee, *J. Am. College Toxicol.*, *2*, 85 (1983).

2. W. T. Stott and P. G. Watanabe, *Drug Metab. Rev.*, *13*, 853 (1982).

3. A. Furst and S. B. Radding, *J. Environ. Sci. Health*, *2*, 103 (1984).

4. F. W. Sunderman, Jr., *Biol. Trace Element Res.*, *1*, 63 (1979).

5. J. D. Heck and M. Costa, *Biol. Trace Element Res.*, *4*, 319 (1982).

6. J. D. Heck and M. Costa, *Biol. Trace Element Res.*, *4*, 71 (1982).

7. N. T. Christie and M. Costa, *Biol. Trace Element Res.*, *6*, 139 (1984).

8. K. Hansen and R. M. Stern, *J. Am. College Toxicol.*, *3*, 381 (1984).

9. A. S. Wright, *Mutat. Res.*, *75*, 215 (1980).

10. H. V. Malling, *Mutat. Res.*, *13*, 425 (1971).

11. J. A. Holme, E. Soderlund, and E. Dybing, *Acta Pharmacol. Toxicol.*, *52*, 348 (1983).

12. J. C. Mirsalis, C. K. Tyson, and B. E. Butterworth, *Environ. Mutagen.*, *4*, 553 (1982).

13. J. E. Lee, R. B. Ciccarelli, and K. W. Jennette, *Biochemistry*, *21*, 771 (1982).

14. J. E. Gruber and K. W. Jennette, *Biochem. Biophys. Res. Commun.*, *82*, 700 (1978).

15. H. J. Phillips, in *Tissue Culture Methods and Applications* (P. R. Kruse and M. K. Patterson, eds.), Academic Press, New York, 1973, pp. 406-408.

16. M. D. Waters, D. E. Gardner, C. Aranyi, and D. L. Coffin, *Environ. Res.*, *9*, 32 (1975).

17. J. M. Zarling, M. McKeough, and F. H. Bach, *Transplantation*, *21*, 468 (1976).

18. L. Muller and F. K. Ohnesorge, *Toxicology*, *31*, 297 (1984).

19. M. A. Sirover, D. K. Dube, and L. A. Loeb, *J. Biol. Chem.*, *254*, 107 (1979).

20. K. W. Jennette, *Environ. Health Perspect.*, *40*, 233 (1981).

21. M. A. Sirover and L. A. Loeb, *Science*, *194*, 1434 (1976).

22. F. O. Brady, *Trends Biochem. Sci.*, *7*, 143 (1982).

23. M. Webb (ed.), *The Chemistry, Biochemistry and Biology of Cadmium*, Elsevier/North Holland, New York, 1979.

24. J. H. R. Kägi and M. Nordberg (eds.), *Metallothionein*, Birkhäuser Verlag, Basel, 1979.

25. F. N. Kotsonis and C. D. Klaassen, *Toxicol. Appl. Pharmacol.*, *42*, 583 (1977).

26. A. J. Zelazowski and J. K. Piotrowski, *Acta Biochim. Polon.*, *24*, 97 (1977).

27. S. Onosaka and M. G. Cherian, *Toxicol. Appl. Pharmacol.*, *63*, 270 (1982).

28. R. W. Olafson and R. G. Sim, *Anal. Biochem.*, *100*, 343 (1979).

29. C. C. Chang, R. J. Vander Mallie, and J. S. Garvey, *Toxicol. Appl. Pharmacol.*, *55*, 94 (1980).

30. C. Tohyama and Z. A. Shaikh, *Fund. Appl. Toxicol.*, *1*, 1 (1981).

31. S. R. Patierno, N. R. Pellis, R. M. Evans, and M. Costa, *Life Sci.*, *32*, 1629 (1983).

32. D. M. Durnam, F. Perrin, F. Gannon, and R. D. Palmiter, *Proc. Nat. Acad. Sci. U.S.A.*, *77*, 6511 (1980).

33. D. M. Durnam and R. D. Palmiter, *Mol. Cell Biol.*, *4*, 484 (1984).

34. D. M. Durnam and R. D. Palmiter, *J. Biol. Chem.*, *256*, 5712 (1981).

35. M. Karin and H. R. Herschman, *Eur. J. Biochem.*, *113*, 267 (1981).

36. K. Hayashi, H. Fujiki, and T. Sugimura, *Cancer Res.*, *43*, 5433 (1983).

37. R. D. Andersen and U. Weser, *Biochem. J.*, *175*, 841 (1978).

38. D. Holt, L. Magos, and M. Webb, *Chem. Biol. Interact.*, *32*, 125 (1980).

39. F. A. Day, A. E. Funk, and F. O. Brady, *Chem. Biol. Interact.*, *50*, 159 (1984).

40. G. L. Eichhorn, in *Metal Ions in Genetic Information Transfer* (G. L. Eichhorn and L. G. Marzilli, eds.), Elsevier/North Holland, New York, 1981, pp. 1-46.

41. L. G. Marzilli, in *Metal Ions in Genetic Information Transfer* (G. L. Eichhorn and L. G. Marzilli, eds.), Elsevier/North Holland, New York, 1981, pp. 47-86.

42. M. P. Waalkes and L. A. Poirier, *Toxicol. Appl. Pharmacol.*, *75*, 539 (1984).

43. R. A. Zakour, T. A. Kunkel, and L. A. Loeb, *Environ. Health Perspect.*, *40*, 197 (1981).

44. M. A. Sirover and L. A. Loeb, *Proc. Nat. Acad. Sci. U.S.A.*, *73*, 2331 (1976).

45. D. Brusick, in *Principles and Methods of Toxicology* (A. W. Hayes, ed.), Raven Press, New York, 1982, pp. 223-272.

46. B. N. Ames, J. McCann, and E. Yamasaki, *Mutat. Res.*, *31*, 347 (1975).

47. D. M. Maron and B. N. Ames, *Mutat. Res.*, *113*, 173 (1983).

48. T. Kada, K. Hirano, and Y. Shirasu, in *Chemical Mutagens* (A. Hollander, ed.), Plenum Press, New York, 1980, pp. 149-173.

49. P. Moreau, A. Bailone, and R. Devoret, *Proc. Nat. Acad. Sci. U.S.A.*, *73*, 3700 (1976).

50. L. Poirier and F. J. de Serres, *J. Nat. Cancer Inst.*, *62*, 919 (1979).

51. G. R. Mohn, *Mutat. Res.*, *87*, 191 (1981).

52. T. G. Rossman, *Environ. Health Perspect.*, *40*, 189 (1981).

53. M. A. Sirover, *Environ. Health Perspect.*, *40*, 163 (1981).

54. N. Kanematsu, M. Hara, and T. Kada, *Mutat. Res.*, *77*, 109 (1980).

55. H. Nishioka, *Mutat. Res.*, *31*, 185 (1975).

56. R. A. Zakour and B. W. Glickman, *Mutat. Res.*, *126*, 9 (1984).

57. T. G. Rossman, M. Molina, and L. W. Meyer, *Environ. Mutagen.*, *6*, 59 (1984).

58. H. Rosenkrantz, B. Sutter, and W. T. Speck, *Mutat. Res.*, *41*, 61 (1976).

59. R. S. U. Baker, A. M. Bonin, A. Arlauskas, R. K. Tandon, P. T. Crisp, and J. Ellis, *Mutat. Res.*, *138*, 127 (1984).

60. T. G. Rossman, *Mutat. Res.*, *91*, 207 (1981).

61. J. S. Dubins and J. M. LaVelle, *Environ. Mutagen.*, *7*, 20 (1985).

62. J. M. LaVelle, *Environ. Mutagen.*, *7*, 46 (1985).

63. R. Mandel and H. J. P. Ryser, *Mutat. Res.*, *138*, 9 (1984).

64. H. Mochizuki and T. Kada, *Mutat. Res.*, *95*, 145 (1982).

65. T. Kada and N. Kanematsu, *Proc. Jpn. Acad.*, *54*, 234 (1978).

66. T. Kada, T. Inoue, A. Yokoiyama, and L. B. Russel, in *Radiation Research, Proceedings of the Sixth International Congress of Radiation Research* (S. Okada, H. Imamura, T. Terashima, and H. Yamaguchi, eds.), Japanese Association for Radiation Research, Tokyo, 1979, pp. 711-720.

67. T. Inoue, Y. Ohta, Y. Sadaie, and T. Kada, *Mutat. Res.*, *91*, 41 (1981).

68. J. F. Hatcher and G. T. Bryan, *Mutat. Res.*, *148*, 13 (1985).

69. M. M. Jacobs, T. S. Matney, and A. C. Griffin, *Cancer Lett.*, *2*, 319 (1977).

70. C. Ip, *Cancer Res.*, *41*, 2683 (1983).

71. M. P. Abbracchio, R. M. Evans, J. D. Heck, O. Cantoni, and M. Costa, *Biol. Trace Element Res.*, *4*, 289 (1982).

72. J. D. Heck and M. Costa, *Cancer Res.*, *43*, 5652 (1983).

73. M. P. Abbracchio, J. Simmons-Hansen, and M. Costa, *J. Toxicol. Environ. Health*, *9*, 663 (1982).

74. P. B. Harnett, S. H. Robison, D. E. Swartzendruber, and M. Costa, *Toxicol. Appl. Pharmacol.*, *64*, 20 (1982).

75. M. Costa, O. Cantoni, M. deMars, and D. E. Swartzendruber, *Res. Commun. Chem. Pathol. Pharmacol.*, *38*, 405 (1982).

76. M. Costa, M. P. Abbracchio, and J. Simmons-Hansen, *Toxicol. Appl. Pharmacol.*, *60*, 313 (1981).

77. M. Costa and M. M. Mollenhauer, *Cancer Res.*, *40*, 2688 (1980).

78. M. Costa and M. M. Mollenhauer, *Science*, *209*, 515 (1980).

79. F. W. Sunderman, Jr., in *Environmental Carcinogenesis* (P. Emmelot and E. Kriek, eds.), Elsevier/North Holland, Amsterdam, 1979, pp. 165-192.

80. E. Nieboer and D. H. S. Richardson, *Environ. Pollut. Ser. B*, *1*, 3 (1980).

81. C. E. Hildebrand, R. A. Tobey, E. W. Campbell, and M. D. Enger, *Exp. Cell Res.*, *124*, 237 (1979).

82. M. D. Enger, L. B. Rall, R. A. Walters, and C. E. Hildebrand, *Biochem. Biophys. Res. Commun.*, *93*, 343 (1980).

83. M. D. Enger, L. T. Ferzoco, R. A. Tobey, and C. E. Hildebrand, *J. Toxicol. Environ. Health*, *7*, 675 (1981).

84. K. W. Kohn, R. A. G. Ewig, L. C. Erickson, and L. A. Zwelling, in *DNA Repair: A Laboratory Manual of Research Techniques* (P. C. Hanawalt, ed.), Marcel Dekker, New York, 1981, pp. 379-401.

85. R. B. Ciccarelli, T. H. Hampton, and K. W. Jennette, *Cancer Lett., 12,* 349 (1981).

86. R. B. Ciccarelli and K. E. Wetterhahn, *Cancer Res., 42,* 3544 (1982).

87. M. J. Tsapakos, T. H. Hampton, and K. W. Jennette, *J. Biol. Chem., 256,* 3623 (1981).

88. O. Cantoni and M. Costa, *Mol. Pharmacol., 24,* 81 (1983).

89. N. T. Christie and M. Costa, *Biol. Trace Element Res., 5,* 55 (1983).

90. N. T. Christie, O. Cantoni, R. M. Evans, R. E. Meyn, and M. Costa, *Biochem. Pharmacol., 33,* 1661 (1984).

91. T. Ochi, T. Ishiguro, and M. Ohsawa, *Mutat. Res., 122,* 169 (1983).

92. T. Ochi, M. Takayanagi, and M. Ohsawa, *Toxicol. Lett., 18,* 177 (1983).

93. O. Cantoni, N. T. Christie, A. Swan, D. B. Drath, and M. Costa, *Mol. Pharmacol., 26,* 360 (1984).

94. S. H. Robison and M. Costa, *Cancer Lett., 15,* 35 (1982).

95. S. H. Robison, O. Cantoni, and M. Costa, *Carcinogenesis, 6,* 657 (1982).

96. M. Costa, J. D. Heck, and S. H. Robison, *Cancer Res., 42,* 2757 (1982).

97. S. H. Robison, O. Cantoni, J. D. Heck, and M. Costa, *Cancer Lett., 17,* 273 (1983).

98. A. D. Mitchell, D. A. Casciano, M. L. Meltz, D. E. Robinson, R. H. C. San, G. M. Williams, and E. S. Von Halle, *Mutat. Res., 123,* 363 (1983).

99. J. J. Yunis, *Science, 221,* 227 (1983).

100. S. M. Galloway, A. D. Bloom, M. Resnick, B. H. Margolin, F. Nakamura, P. Archer, and E. Zeiger, *Environ. Mutagen., 7,* 1 (1985).

101. R. J. Preston, W. Au, M. A. Bender, J. G. Brewen, A. V. Carrano, J. A. Heddle, A. F. McFee, S. Wolff, and J. S. Wassom, *Mutat. Res., 87,* 143 (1981).

102. P. E. Perry and S. Wolff, *Nature, 251,* 156 (1974).

103. S. A. Latt, R. R. Schreck, K. S. Loveday, and C. F. Shuler, in *Cytogenetic Assays of Environmental Mutagens* (T. C. Hsu, ed.), Allanheld, Osmun, Totowa, New Jersey, 1982, pp. 29-80.

104. H. Vainio and M. Sorsa, *Environ. Health Perspect.*, *40*, 173 (1981).

105. H. Ohno, F. Hanaoka, and M. Yamada, *Mutat. Res.*, *104*, 141 (1982).

106. P. Sen and M. Costa, *Cancer Res.*, *45*, 2320 (1985).

107. T. C. Hsu, *Genetics*, *79*, 137 (1975).

108. A. W. Hsie, D. A. Casciano, D. B. Couch, D. F. Krahn, J. P. O'Neill, and B. L. Whitfield, *Mutat. Res.*, *86*, 193 (1981).

109. D. Clive, K. O. Johnson, J. F. S. Spector, A. G. Batson, and M. M. M. Brown, *Mutat. Res.*, *59*, 61 (1979).

110. P. Howard-Flanders, *Mutat. Res.*, *86*, 307 (1981).

111. M. O. Bradley, B. Bhuyan, M. C. Francis, R. Langenbach, A. Peterson, and E. Huberman, *Mutat. Res.*, *87*, 81 (1981).

112. T. J. Oberly, B. J. Bewsey, and G. S. Probst, *Mutat. Res.*, *125*, 291 (1984).

113. J. W. Greiner, J. A. DiPaolo, and C. H. Evans, *Cancer Res.*, *43*, 273 (1983).

114. M. A. Cifone, J. Fisher, and B. Myhr, *Environ. Mutagen.*, *6*, 423 (1984).

115. M. A. Cifone, *Environ. Mutagen.*, *7*, 27 (1984).

116. D. Brusick, in *Proceedings of the Toxicology Forum 1984, Winter Meeting, February 20-22, 1984,* Toxicology Forum, Washington, D.C., 1984, pp. 139-152.

117. S. M. Galloway, C. L. Bean, M. A. Armstrong, D. Deasey, A. Kraynak, and M. O. Bradley, *Environ. Mutagen.*, *7*, 48 (1984).

118. A. L. Meyer, *Mutat. Res.*, *115*, 323 (1983).

119. R. J. Pienta, J. A. Poiley, and W. E. Lebherz III, *Int. J. Cancer*, *19*, 642 (1977).

120. C. A. Reznikoff, J. S. Bertram, D. W. Brankow, and C. Heidelberger, *Cancer Res.*, *33*, 3239 (1973).

121. J. A. DiPaolo, K. Takano, and N. C. Popescu, *Cancer Res.*, *32*, 2686 (1972).

122. B. C. Casto, J. Meyers, and J. A. DiPaolo, *Cancer Res.*, *39*, 193 (1979).

123. V. C. Dunkel, R. J. Pienta, A. Sivak, and K. A. Traul, *J. Nat. Cancer Inst.*, *67*, 1303 (1981).

124. C. Heidelberger, A. E. Freeman, R. J. Pienta, A. Sivak, J. S. Bertram, B. C. Casto, V. C. Dunkel, M. W. Francis, T. Kakunaga, J. B. Little, and L. M. Schechtman, *Mutat. Res.*, *114*, 283 (1983).

125. F. W. Sunderman, Jr., and R. M. Maenza, *Res. Commun. Chem. Pathol. Pharmacol.*, *14*, 319 (1976).

126. J. D. Heck and M. Costa, *Cancer Lett.*, *15*, 19 (1982).

127. M. P. Abbracchio, J. D. Heck, and M. Costa, *Carcinogenesis*, *3*, 175 (1982).

11

Some Problems Encountered in the Analysis of Biological Materials for Toxic Trace Elements

Hans G. Seiler
Institute of Inorganic Chemistry
University of Basel
CH-4056 Basel, Switzerland

1. GENERAL ASPECTS OF TRACE ELEMENT ANALYSIS

During the last 25 years trace element contents in biological systems have been receiving increasing interest. Though specialized scientists had recognized already many years ago the important role of some trace elements in the functioning of several biochemical

processes, the general interest of nearly all disciplines of biological and medical sciences, as well as of geography and agronomics,
has only recently been attracted by trace elements.

There are several reasons for this development. On the one
hand the high effectiveness of certain metal ions in biological
systems—toxic and essential ones—has become common knowledge; on
the other hand, during the last two decades great progress has been
made in developing new analytical methods and in perfecting the
related instrumentation. These developments were directed to enhance
sensitivity, precision, and reproducibility, but also to facilitate
analytical determinations. In former times complicated analytical
procedures, such as the determination of elements at very low concentrations in a complex matrix, were restricted to specialized
analysts. Today, encouraged by new methods and sophisticated instrumentation, many people are discovering themselves capable of carrying
out such determinations, sometimes even without any knowledge of the
chemistry of the elements to be determined or the matrices containing
these elements.

Incited by many reports of health hazards, the public is now
demanding information on toxic elements in food and the environment.
This demand is reflected in the increasing number of published data
and methods on so-called toxic elements compared with those on the
essential elements. The public interest in these data on toxic
elements is of consequence to the analyst. Only a few years ago
results of analytical measurements describing the detrimental action
of toxic elements were mainly of scientific interest; today political
and economical decisions, affecting nearly every domain of the human
society, are also influenced by them: hence analytical chemistry has
become an integral part of modern society. Consequently, the responsibility of all those carrying out such analyses has become much
greater, and all processes involving the accumulation of data must
be undertaken with the utmost conscientiousness.

As indicated, the development of methods for trace element
analysis was directed above all to increase sensitivity, precision,

specificity, and reproducibility, but also to facilitate measurements. Today mainly three methods are used for these kinds of
analyses: atomic absorption spectrometry (AAS), stripping voltammetry (SV), and neutron activation analysis (NAA). By interlabora-
tory comparative analyses of certified reference materials, the
random and systematic errors associated with sample preparation and
subsequent analytical determinations can be recognized and overcome.
However, the reliability and accuracy of analytical data for biological materials do not only depend on the analytical laboratory work;
they are influenced by all procedures that the specimen and the
sample derived from it undergo, from their collection to the statistical evaluation of the measurements. Unfortunately, in most cases
the analyst has only limited influence, if any, on the manipulations
which occur before the specimen arrives at the laboratory. In all
these initial stages the true content of an element may be altered
by contamination or loss. Whereas formerly most of the attention
was given to avoiding losses, today analysts are aware of the much
greater problems arising from contaminations. Thus today the values
for normal levels of several elements in human blood must be revised;
the corresponding concentrations are now considered too high because
the samples had been contaminated with the elements in question [1-4].
Therefore the early warning issued by Thiers in 1957 is still valid:
"Unless the complete history of any sample is known with certainty,
the analyst is well advised not to spend his time in analysing it"
[5].

In order to elucidate and estimate the possible errors occurring during the several stages of an analysis for trace elements in
biological materials, these stages should be subdivided into two
phases: (a) the preanalytical phase, consisting of sampling, mail-
ing, storage, homogenization, and aliquotation, and (b) the analytical
phase, comprising sample treatment, measurement, and statistical
evaluation. Whereas errors in the analytical phase can be recognized,
evaluated, and also largely eliminated by controls using certified
biological reference materials and statistical quality controls of

the methods employed, there is no method of control and correction
for the preanalytical phase. All alterations of the biological
material during this phase may modify the result of the measurements
in vitro; these are the so-called "interference factors" [6]. They
can only be minimized by scrupulous standardization of all manipulations, observing all possible sources of errors.

2. CHOICE OF IMPLEMENTS AND CHEMICALS

All materials and chemicals coming into contact with the specimen or
the sample derived from it should be selected with regard to possible
contaminations, loss, and interfering interactions. For this their
exact composition, mode of fabrication, and behavior should be well
known. But even the best information does not release the analyst
from testing all materials and chemicals for their suitability, not
only with respect to the element to be determined but also with
respect to the biological material. Every new lot of materials and
chemicals must be retested before use.

All chemicals, especially those needed in larger quantities,
must be of the highest purity available. For determinations at very
low concentrations (ppt range) even reagents of very high purity must
be further purified. Therefore the selection of chemicals should be
made not only with respect to their chemical properties but also with
regard to the ease and quality of their purification. Hence acids and
bases which are distillable at lower temperatures are preferred, for
example, HNO_3, HCl, and NH_3. Techniques for preparing exceptionally
high-purity acids have been reported by several authors [7-9].

Chemicals and solutions of reagents are preferably stored in
containers fitted with dispensers in order to prevent contamination
by opening the bottles and inserting pipettes. Generally, polyolefin,
and above all polyethylene, containers have proved to be adequate for
specimen collection and the storage of biological materials. But
polyethylene and polypropylene are collective names for a great variety of materials with different characteristics. Therefore polymers

used for containers must be tested for suitability. Special attention must be given to colored stoppers, because pigments may contain
elements of interest.

Independent of the particular polymer, all containers must be
carefully cleaned before use, because during manufacturing their
surface can be contaminated with different metals. The lower the
expected concentration in the specimen, the more thorough the cleaning must be. In most cases soaking the containers with diluted HNO_3
(10%) during several days, renewing the acid every day, followed by
a thorough rinsing with high-purity water is sufficient [10]. If
the biological material may undergo undesired changes by acids, the
containers must be thoroughly soaked with high-purity water, because
acids are able to penetrate polymers and are only slowly released.
High concentrations of oxidizing acids should be avoided for cleaning,
as these may create active adsorption sites in the polymers, causing
possible losses of trace elements from the stored specimen [10].
Polyolefin bottles destined to store specimens (e.g., water) for the
subsequent determination of mercury should previously be treated with
chloroform and aqua regia vapor; in this way losses of mercury due to
reactions with additives in the polyolefin plastic can be avoided [11].

Constructional materials for the apparatus, vessels, and tools for
specimen collection and sample preparation must, of course, not be contaminating or absorbing and must be resistant to specimens and chemicals.
The surface of the materials must be smooth and easy to clean. With
this in mind, pure silica shows the best properties, except for alkaline melts. Sometimes Teflon is recommended, but it should be kept
in mind that Teflon is a sintered product with grain boundaries which
may give rise to loss of elements by absorption, especially at elevated temperatures. Besides this, Teflon shows a tendency for memory
effects; that is, reuse of a vessel may give rise to contamination of
a sample by the release of previously absorbed ions. Glassware of
high silica content like Pyrex may also be used but must previously
be leached with boiling concentrated HNO_3.

Metal tools should never contain elements to be determined.
Their surface must be smooth and free of oxide films, unless these

are very thin, compact, and resistant to the chemicals used. Metal tools should never undergo reactions with the metal ions of interest or with the biological material. The contact of biological material with any metal for a long period must be avoided. Knives for cutting or mixing, scalpels, needles for syringes, and so on, are preferably made of pure metals that have, as far as we now know, no biological significance—for example, titanium or tantalum—or of high-purity plastics or silica.

In developing the methods of analysis—from specimen collection to measurement—it is important to ensure as little contact as possible between the specimens or samples and the surfaces of the used materials. The equipment should be of such design that it allows several stages of the procedures to be carried out in the same apparatus or vessel.

Special attention must be given to the cleanliness of the laboratories, especially to the air. When laboratories are located near roads with heavy traffic or in the neighborhood of power plants, industries, or incinerator plants, dust may become a considerable source of contamination, sometimes even prohibiting trace element analysis. Already very small particles of dust may cause contaminations of the specimens and samples which exceed the true content of the element by orders of magnitude [12,13]. In these cases all laboratory work should be done in a clean-air room or within a laminar flow work station fitted with a high-efficiency particulate air filter. This latter precaution prevents contaminations with particulate matter but does not eliminate gaseous substances, for example, mercury or volatile compounds of other elements. When using laminar flow work stations, one must be aware that installation of apparatus and minipulations inside the work station may disrupt the laminar flow and thus generate turbulences by which dust from the outside can penetrate. Therefore after every change of the installation the efficacy of the work station should be tested [7].

The risk of contamination with a specific element is correlated with its content in the airborne dust. Typical data for dust in a

laboratory in an urban environment have been published by Sansoni
and Iyengar [14]: Al, 3000 ppm; Cd, 3 ppm; Co, 9 ppm; Cr, 39 ppm;
Cu, 213 ppm; Mn, 116 ppm; Ni, 70 ppm; Pb, 2150 ppm; V, 259 ppm; Zn,
1640 ppm. This would mean that a specimen of 1 g containing 0.1 mg
of such dust would show a value for lead of 215 ppb originating only
from contamination. The analyst himself can also be a source of
contamination via his clothes, hairs, skin, and breath.

     Generally it can be said that the smaller the specimen and the
lower the trace element content, the greater the efforts to prevent
contaminations or losses must be.

3. SPECIMEN COLLECTION

Specimen collection and sampling are the most important but also the
most critical steps in every analysis. The general analytical prin-
ciple that no result of an analysis can be better than the sample
from which it was obtained is especially valid for trace element
analyses in biological materials. The specimen and the sample
derived from it must exactly reflect the properties of the matrix.
As such matter is very complex, the planning of this kind of analysis,
and above all the collection of the specimens, should only be undertaken in close interdisciplinary collaboration between a scientist in
the field of specialty and the analyst. Unfortunately, in the past
analysts without sufficient knowledge about biological systems, as
well as biologists and physicians without analytical training, produced a great number of data of dubious value on trace elements in
biological materials.

     It is obvious that a major requirement for trace element analy-
sis is that the specimens be taken under noncontaminating conditions
and thus avoiding all kind of losses. This requirement is not easy
to fulfill, as sometimes specimen collection must be effected under
difficult conditions. In any case, specimen containers and collection implements should be prepared and made available by the analyst

whenever specially prepared specimen-collection systems are not disposable.

It is recommended that the analyst or a person well instructed by him execute the specimen collection in order to be able to recognize and prevent possible contaminations and losses.

For ethical and professional reasons, the sampling of human blood and tissue can only be carried out by medical staff. These persons must be well informed about the analytical requirements, especially with respect to tools (needles for syringes, scalpels, etc.), disinfectants, and preservatives. In this respect, it is important to remember that sterility does not imply the absence of contaminating elements.

The risks for inaccurately evaluating the composition of a specimen, as well as for contaminations and losses, are different for liquid and solid specimens.

3.1. Liquid Specimens

Liquid specimens of biological materials—mostly body fluids or the juice of plants—may be considered nearly homogeneous with respect to their composition at the moment of collection. Therefore the size of the specimen will not affect the accuracy of the analysis. The quantity to be taken is above all determined by the concentration of the trace element to be measured and the sensitivity of the applied analytical method. But, as biological fluids may undergo different alterations (coagulation, precipitation, etc.) soon after collection, preservatives must often be added immediately after collection. As trace element concentrations in body fluids are normally low—except in the case of intoxication—special care must be given to prevent exogenous contaminations. By adequately cleaning the subject from which the specimen is collected, using specially prepared tools and containers, and excluding dust, one can minimize contamination of the specimen.

In the specimen collection of body fluids, the physiological facts must also be considered [15]. Body fluids may be "internally contaminated" by nutrition. Heydorn et al. [16] found that after the intake of food which consisted mainly of a fillet of plaice, the arsenic level in blood serum rose by more than one order of magnitude. Even after 12 hr, the normal fasting period before a blood specimen is taken, the concentration was still elevated. This demonstrates that erroneous results can be obtained by the entry of an element into the body via nutrition, if the specimen collection is made before a return to the normal level.

3.1.1. Specimen Collection of Blood

In order to prevent exogenous contaminations, the subject from which the specimen is to be collected should preferably undergo a whole-body washing. At least the forearm must be thoroughly washed with soap and water, because particles of inorganic substances may strongly adhere to the skin.

The position of the subject during collection influences the composition of the specimen. When a person changes from an upright to a recumbent position, extracellular fluid flows into the blood vessels, causing a dilution of the protein in the blood plasma. Consequently, the concentration of protein-bound elements (e.g., Zn and Se) is decreased. This change, which can be as great as 20%, may give rise to erroneous interpretations, especially if data from subjects whose blood was collected in a recumbent position are compared with those from a group which had been upright during collection [17].

Disinfectants must be carefully controlled for contaminating elements. Note: If mercury is to be determined, the arm should never be disinfected with mercury-containing agents (e.g., Merfen and Mercurochrome).

Blood specimens should be collected using disposable needles and syringes. If nickel, chromium, manganese, or cobalt are to be

measured, the usual needles made from stainless steel must be
replaced by needles of plastic, silver, or titanium.

    The blood is transferred from the syringe to disposable con-
tainers of polyethylene which have been washed with 1 M HNO_3 and
thoroughly rinsed with high-purity water. As inhomogeneity, caused
by coagulation, would be a serious source of error, especially if
only small quantities of samples are used for the determination,
adequate amounts of anticoagulants must be added immediately after
the specimen has been collected. The anticoagulant must be free
of contaminating elements. Thus heparin cannot be used, its contents of trace elements being too high. Potassium-EDTA has proved
to be particularly suitable for this purpose.

    During transport, storage, and sometimes even sample prepara-
tion, the blood remains in the aforementioned polyethylene tubes.
Today specially prepared tubes already containing an anticoagulant
are commercially available (e.g., Monovette and Vacutainer) [6].

3.1.2. Specimen Collection of Urine

Whereas the composition of blood may be considered nearly constant
in regard to its main constituents, the situation is quite different
in the case of urine. The composition of urine, a final product of
the metabolism, varies over a wide range, not only with respect to
the main constituents but also with respect to the trace element
content. It is influenced by physiological factors, as well as by
nutrition, state of health, and physical and psychical stress. The
composition varies not only from one individual to another but also
within the different excretions of the same individual. Therefore
the collection of all excretions during a period of 24 hr would be
the desirable specimen. However, this is rather difficult to realize,
except for hospitalized persons. Collected excretions should be well
defined with respect to time, for example, the second excretion in
the morning [18] or at the end of a working shift [6]. Calculations
from one spontaneous excretion to the total excretion of a day are
usually made by determining the creatinine content. The validity

of these back-calculations with respect to trace element content is
still subject to discussion.

The main risk for exogenous contamination in the specimen
collection of urine is the subject itself; therefore a whole-body
washing prior to the collection is indispensable. Clothes which
could be a source of contamination should be changed. The urine is
usually excreted into plastic containers with large openings. These
are not defined, as special containers are not yet available. The
containers must be scrupulously cleaned and rinsed with high-purity
water. Immediately after collection an aliquot of the urine is
transferred to a specimen tube. In order to prevent sedimentation
and thus inhomogeneity of the specimen, the urine is generally acidified. This acidification is also important for its preservation from
bacterial decomposition. However, there are exceptions: If the
specimen collection takes place in the absence of persons aware of
the analytical requirements, the addition of acid may be inaccurate
and contaminations cannot be excluded. In these cases the specimens
are preferably acidified at their arrival in the analytical laboratory.

Often the addition of mineral acid (HNO_3, HCl, and $HClO_4$, all
of the highest purity grade) to guarantee a pH of not more than 2
is recommended [18]; however, the pH should also be selected in
accordance with the requirements of the subsequent analytical steps.
Extraction procedures as used for the isolation or enrichment of
trace elements often require pH values between 2 and 5. The pH of
highly acidic urine specimens would have to be readjusted with a
base, thus introducing a further source of contamination. These
difficulties can be avoided by the use of acetic acid as a preservative. Years of experience have shown the following procedure to be
very suitable: 1 ml of glacial acetic acid is added per 100 ml of
urine. After acidification, specimens obtained from a great range
of different sources reveal a pH between 3.3 and 4.3. Thus the
urine is both adequately preserved and adjusted to a pH favorable
for many analytical procedures. Beside this, owing to the buffering

effect, the ratio of the quantity of acetic acid to urine is not critical [6]. However, this procedure should not be used for specimens destined for measurement of the total mercury content. These samples must be acidified to a pH of not more than 1 with HNO_3.

Specimens destined for determinations of organometallic compounds and mercury are not to be stored in plastic tubes, as these substances are able to penetrate the plastic and thus give rise to considerable losses. In these cases glass tubes are preferred.

3.2. Specimen Collection of Tissues

In contrast to liquids, tissues cannot be considered homogeneous, neither to their trace element content nor to the composition of the biological matrix. The distribution of trace elements depends on both the element and the specific organ. Therefore the accuracy of the analysis is closely correlated with the kind of specimen and its size. Generally the organs in which a specific element is accumulated are the preferred materials, as the content in these organs is significantly elevated and therefore the risk of obtaining erroneous results due to contamination is reduced [18].

Because the distribution of trace elements within tissue is never uniform, every specimen should be as large as possible. Whenever practical, the whole organ of interest should be taken into the specimen. It was found, for example, that the cadmium content of human kidneys varies over a wide range within each kidney: It is highest in the cortex, decreases in the medulla, and is lowest in the papillae; however, there are also variations within the cortex that have been observed [19]. Therefore analytical results obtained from small parts of organs, for example, material from a biopsy, should, in addition, be specified by histological investigations. Problems concerning the accuracy of analytical results may also arise from the inhomogeneity of the biological matrix. Thus the presence of adipose tissue, which generally has low concentrations

of trace elements, or variations of the water content in the tissue
may falsify the results with respect to the examined organ [15].

Normally no preservatives are added to tissue specimens. The
specimens are placed in adequately cleaned plastic containers and
cooled as soon as possible in order to prevent deterioration and
losses of water. The weight of specimens and samples, to which the
contents of trace elements are related, is one of the main difficulties in the analysis of biological materials. The so-called
fresh weight is delicate to define after specimen collection. This
can be overcome, at least to a certain extent, by using tared specimen containers with tight stoppers. Thus losses of water are minimal
and the true weight of the specimen can be determined.

3.2.1.   Specimen Collection of Animal
             and Human Tissues

Collections of specimens of animal or human tissue normally take
place in laboratories or hospitals, where it is possible to create
adequate noncontaminating conditions. Because in most tissues the
concentrations of trace elements are significantly elevated compared
with those in body fluids, the risks for erroneous results due to
exogenous contaminations are decreased. In spite of this, all precautions concerning contaminating tools and chemicals must be scrupulously observed [20]. The surface of the body from which the specimen
is excised must be thoroughly cleaned. Material excised near an organ
with a high content of an element can be contaminated by this if the
operator does not pay great attention. The outflow of intracellular
liquid during excision is a frequent source of losses. Material from
an autopsy must be taken as soon as possible postmortem, because
deterioration of the tissue starts immediately after death, altering
the contents of the elements [21,22].

After collection animal and human tissues must be immediately
put into a specimen container and cooled. Preservatives such as
formalin, usually used in medicine, are not suitable, for the risks
of contamination are too high.

3.2.2. *Specimen Collection of Tissues from Plants*

Mostly whole plants or parts of them must be collected in the field, where they were exposed to the environment. Particles of dust or soil may adhere strongly to their surface, especially if leaves, stems, or fruits are covered with tiny hairs or wax. These particles create a high risk for contamination. For specimen collection small plants should be taken in their entirety into the laboratory and decontaminated by a thorough rinse with distilled water before being cut, in order to prevent losses of extravascular juice and contamination by penetrating wash water. Plants should never be immersed in water for long periods because water may penetrate the plant or elements inherent to the plant may be extracted. The use of detergents or organic solvents for decontamination is not recommended, as they can cause the biological matrix to deteriorate. After rinsing, the specimens are wiped dry with filter paper, without pressing or rubbing. The time elapsed between collection in the field and the laboratory work-up should be as short as possible to avoid loss of water. Plants with a high water content must be transported in a cooled tank.

After decontamination, the parts or organs of interest are cut directly into the specimen containers with the use of very sharp tools of noncontaminating material (e.g., knives or scissors made of titanium). If the trace element to be determined is of very low concentration or has a high ubiquitous occurrence, such as in the case of aluminum, all laboratory work should be done in a laminar flow work station with efficient air filters or in a special "clean room." Metal-free plastic gloves should be worn.

Parts of large plants, which cannot be taken as a whole into the laboratory, are difficult to decontaminate before cutting. Decontamination after cutting may cause the juice of the plant to leak out at the surface of the cut and water to penetrate, both events affecting the true composition of the specimen. This problem can be overcome by collecting rather large parts of the plant in the field and decontaminating them in the laboratory under adequate

conditions. The specimen should consist of only those parts that
are at a sufficient distance from the original surface of cut, so
that interfering events cannot occur.

4. STORAGE, DRYING, AND HOMOGENIZATION

The second important step of the preanalytical phase is the storage.
Seldom can specimens be analyzed immediately after collection; they
need to be stored for a certain period. Often they must be mailed
to the analytical laboratory, because the sophisticated instrumentation required for this kind of analysis is not available everywhere.
Long-term storage is necessary for specimen-banking authentic materials for retrospective and comparative analyses [23,24]. During
storage the specimens may undergo a great variety of alterations by
chemical and bacterial interactions. Plastic containers, which are
necessary for low-temperature storage, do not exclude the penetration
of oxygen and specimens are seldom sterile.

A general precaution taken to reduce these detrimental actions
is to store specimens in the dark at temperatures between +4 and
-196°C. The temperature will depend on the kind of specimen and the
duration of the storage.

Generally, interactions—chemical and bacterial—are favored
by the presence of water. Drying prior to storage is advised in
certain cases, but this process irreversibly alters the biological
matrix, because water and all volatile constituents are eliminated.
The so-called dry weight frequently used for comparisons of results
from different materials and laboratories is as difficult to define
as the fresh weight [18]. The dry weight is influenced by any alterations of the biological material that take place between specimen
collection and drying, the applied temperature, and the kind of biological material. Drying may also influence the sample preparation
procedure, as will be seen later. Losses of elements during drying
may occur owing to the formation of insoluble substances (e.g.,
aluminum oxides) and the evaporation of volatile elements or

compounds of elements. Thus a large portion of inorganic and organic bound mercury is lost by drying. Similar effects can be observed for arsenic and selenium [25,26].

Lyophilization is to be preferred over oven drying, as at low temperatures fewer alterations of the biological material occur and the formation of insoluble substances is also decreased. Losses due to evaporation are also reduced, but not excluded. Contaminations by drying arise mainly from the material of the drying apparatus employed. Thus specimens for subsequent analyses for chromium, nickel, cobalt, or manganese should never be dried in a stainless steel oven; in such cases aluminum is the preferred material [18,27].

Liquid specimens requiring preservatives, such as blood and urine, are preferably stored at +4°C, provided that they will be analyzed within 3 weeks. If a longer storage period is necessary, however, such specimens should be cooled to -20°C [6]. Losses of elements may occur through the formation of insoluble products that are strongly adsorbing to the wall of the container, coprecipitation with the main inorganic and organic constituents of the matrix, the formation of volatile compounds that penetrate the plastic, and also the formation of compounds which are unsuitable for the subsequent analytical procedure. Most of these detrimental effects are diminished or eliminated by the addition of sufficient amounts of preservative.

Analytically, the most critical element is mercury, even in the presence of preservatives. Mercury can be reduced to its elemental form by the biological matrix and in this state is able to penetrate the plastic, giving rise to a great loss [28]. On the other hand, mercury sulfide is a very insoluble compound; if sulfide is formed by degradation of the biological material during storage, great losses of mercury are to be expected, because mercury sulfide adsorbs strongly to the wall of the container. Therefore specimens to be analyzed for mercury should be stored in glass or silica containers, thus excluding subzero storage temperatures, and should be analyzed as soon as possible after collection.

Contaminations during storage are due mainly to the material of the containers (see Sec. 2). Other interfering effects are loss of homogeneity by alteration of the organic constituents and loss of water, which is due to both the ability of water vapor to penetrate the plastic and the low relative humidity within refrigerators. This latter effect influences the true weight of the specimen but is overcome by the use of tared containers.

In the storage of tissue it is important to know whether the entire specimen or an aliquot sample representative of the mean value of the specimen will be submitted for analysis. In the first case no special precautions are needed. After determination of the fresh weight and, if the specimen is dried, the dry weight, the specimen is cooled as soon as possible to -20°C or below. At this temperature bacterial and chemical interactions are largely diminished. During freezing and thawing the structure of the specimen will be altered by the rupture of membranes and outflow of intracellular liquid; however, since the total specimen will be analyzed, the accuracy of the analysis will not be affected.

However, when a representative aliquot is to be taken from the specimen, a homogenization step prior to storage is essential. The structure and composition of the biological material will be changed by both drying and freezing. Homogenization is mostly carried out in mixers with rotating knives. These instruments are one of the main sources of contamination. Often the composition of the different materials from which the mixer is made is unknown. The most contaminating item is the knife, because there will always be some abrasion. Therefore the knives must be made from materials that are of no analytical interest, for example, titanium or tantalum. Losses may occur by adsorption of interesting elements to the surface of the mixer, as well as by evaporation [29]. High-speed mixers give rise to considerable cavitation by which water and volatile organic substances, as well as volatile elements (Hg), can be lost. Cooling during mixing is advised. The efficiency of mixing with respect to homogenization and contamination must be controlled for each new kind

of specimen by analyzing samples of different size. For the homogenization of biological material with a low water content, distilled
water must be added. All homogenized tissue specimens are preferably
lyophilized before storage to better maintain homogeneity and diminish
bacterial interactions.

5.   SAMPLE ALIQUOTATION AND SAMPLING

Sampling is the last step of the preanalytical phase that can possibly
affect the accuracy of the analytical result via interference factors.
It is of decisive importance in all cases where representative aliquots of the specimen must be taken. Its precision depends above all
on the homogeneity of the specimen and the size of the sample. In
general, correctly homogenized and lyophilized tissue specimens do
not undergo changes in their composition with respect to homogeneity
during storage. Possible alterations in the water content can be
determined.

In contrast, liquid specimens may have become inhomogeneous
during storage, in spite of any added preservatives. Trace elements
are often contained in or adsorbed to precipitated particles. Therefore homogeneity must be restored as best as possible. This is most
important for analytical methods which allow the direct measurement
of trace elements in very small sample volumes (10-100 µl) without
prior sample preparation. Besides the stringent homogeneity requirements, one should keep in mind that all pipetting devices are cali-
brated with water. Liquid biological materials differ from water
with respect to density, viscosity, and surface tension. If the
volume of the small sample to be delivered is inaccurately reported,
since the concentrations of trace elements are mostly related to the
volume, the results of the analysis will be erroneous. Therefore
weighing the delivered liquid from a given volume is an indispensable
test. The possible interfering effects cited are not restricted to
small sample volumes; they are also not to be neglected for the rather

large samples usually employed for analytical procedures with prior
sample preparation steps.

Stored blood specimens show all stages of separation. Before
aliquots are removed, they should be warmed to room temperature and
carefully mixed to effect homogenization. A roller-mixer has proved
to be particularly suited for this purpose: The containers are
placed horizontally within the mixer, which rolls them continuously
along their long axes; at the same time they undergo a slow wave
movement of small angle. The two movements are gentle enough to
avoid foaming. After 1 hr of mixing homogeneity is restored [6].

Urine specimens are mostly inhomogeneous after storage, especially when they were frozen. Manifold investigations have been made
on methods of restoring the homogeneity of trace elements, including
shaking, ultrasonic treatment, heating, and centrifugation. Analyses
of urine carried out before and after each treatment determined that
lead, cadmium, mercury, and arsenic (the only elements analyzed) are
partially bound to the sediment. The only method of restoring homogeneity is to shake the specimens vigorously during at least 30 min
by means of a mechanical shaker. Immediately prior to pipetting a
sample, the specimen must be shaken once more by hand. This demonstrates that true homogeneity cannot be achieved, and consequently
the volume of the sample should not be too small. Unfortunately,
there is no disposable apparatus comparable to the blood roller-mixer
to standardize mixing. This lack is a result of the great variety in
the sizes of the containers used for the collection and storage of
urine [6].

6. SAMPLE PREPARATION

Sample preparation comprises every treatment the biological material
must undergo between sampling and definitive determination of the
element of interest. This is the first step of the analytical phase.
Interference factors of unknown effectiveness can now be excluded.
All random and systematic errors occurring in this and the following

step—the determination—can be recognized and overcome by means of statistical quality control [30,31].

The purpose of sample preparation is to bring the sample and the element of interest in a physical and chemical form suitable to the method of determination, eliminate disturbing substances, and in some cases enrich the element of interest. Typical kinds of sample preparation procedures are dilution, solubilization, acid extraction, solvent extraction after complexation, the extraction, filtration, and centrifugation of interfering substances, ion exchange, oxidation, distillation, and mineralization.

The procedure of choice depends on the nature of the biological matrix, the kind and concentration of the element to be determined, and the method used for the determination. All preparation procedures have a certain potentiality for contaminations and losses. Contaminations arise mostly from the employed chemicals, the air in the laboratory, and the implements. Losses are mainly due to improper separation methods, the application of high temperatures, and vessel transfers. These aspects must dominate the selection of a preparation procedure.

Often in the evaluation of the effectiveness of preparation methods for biological materials, a known quantity of a specific element in the form of an inorganic salt is added to the biological matrix in order to determine its degree of recovery. The results of such experiments on so-called "spiked" materials are dubious, as it is unlikely that the added element will bind to the same sites as the element inherent to the biological matrix. Even the oxidation states can differ. Thus considerably different behaviors are to be expected of the spiked and the inherent element during the whole procedures [15]. Therefore so-called "reference materials" of the same or very similar biological material with certified contents of trace elements should be used to test a new method. A limited assortment of these biological reference materials is available from national and international organizations, for example, the National Bureau of Standards (NBS), Washington; the Community Bureau of Reference (BCR),

Brussels; and the International Atomic Energy Agency (IAEA), Vienna.
The number and variety of available biological reference materials
are increasing. These materials should be used not only as controls
in the development of new analytical methods, but also as controls
in routine analytical work [31]. The only method of determination
needing no sample preparation—except for packing the sample into the
irradiation container—is instrumental neutron activation analysis
(INAA). The only contaminations possible would arise from the composition of the irradiation containers, often pure silica ampoules,
and the flame used to seal them. This largely applies also to radio-
chemical neutron activation analysis (RCNAA), because contamination
by the necessary postirradiation chemical treatments does not influence the result. Only those elements which have been bombarded with
neutrons become radioactive, and it is the emitted radiation that is
quantified. Losses in INAA are almost excluded, while those in RCNAA
can be minimized by the addition of comparatively large quantities of
the same inactive elements (carrier) to all processes of chemical
treatment.

In electrothermal atomic absorption spectrometry (ETAAS) small
volumes of liquid are also sometimes submitted for determination
without prior preparation, but this direct method is not comparable
to neutron activation analysis. The sequence of steps of a determination by ETAAS includes drying and ashing the sample; that is,
a preparation procedure takes place within the graphite furnace and
losses of an element by volatilization are possible. Application of
this method should be restricted to spectrometers with Zeeman background compensation and high resolution of the signal in time. In
addition, the use of a platform in the graphite furnace is advisable.
In the presence of only partially ashed biological material high
background signals are produced in the atomization and measuring
steps. This direct method requires great experience, especially in
interpreting the signal. Accuracy must be controlled by the use of
identical or very similar biological reference material. The calibration is normally carried out by the method of standard addition.

6.1. Acid Treatment

The simplest and least-contaminating preparation procedure is to
treat the sample with dilute acid. At pH $\leqslant$ 0.5 most of the trace
elements are released from their original bondings in biological
materials. This procedure is applicable to liquid and solid samples
for subsequent determinations by flame atomic absorption spectrometry
(FAAS) or ETAAS, provided that the method's sensitivity and specificity
are sufficient and the determination is not affected by components of
the matrix that have gone into solution. This acid treatment has the
advantage that disturbing main constituents of liquid samples are
frequently precipitated—for example, proteins—and can be removed
by centrifugation. Many solid samples do not completely dissolve in
dilute acid and therefore the solid residue can also be removed. In
both cases the metal ions are in the aqueous supernatant with a reduced
content of organic substances and in homogeneous solution. The effectiveness of the procedure must always be tested with respect to the
specific element as well as to the particular biological material.
Besides, one must consider the fact that in FAAS dissolved organic
substances may contribute to the fuel content of the flame and thus
change its temperature. This would lead to erroneous results if the
calibration is made using aqueous solutions of a salt of the element.
Calibration should be made with standard additions of known quantities
of the element to the biological matrix and carrying out the whole
procedure.

This preparation method is used for the ETAAS determination of
cadmium in whole blood [32]. After the addition of a solubilizer
(Triton X-100) the blood is deproteinized with 1 M HNO_3 and the solid
constituents are removed by centrifugation. The cadmium concentration is determined in the supernatant. The standard addition procedure is used for calibration. The risks of contaminations are low,
as only small quantities of reagents are needed. The accuracy of
the procedure has been confirmed by comparison with independent
analytical methods [33].

6.2. Complex Formation, Extraction, and Enrichment

If concentration of the trace element or the sensitivity of the method of determination is too low, it is often necessary to enrich the element. Enrichments are generally made by complex formation of the element with a strong ligand followed by extraction of the complex into a solvent or solvent mixture immiscible with water.

Often enrichment is preceded by acid treatment of the sample and elimination of the insoluble components. The resulting homogenous solution can then be buffered to an appropriate pH. The complexing agent must form very stable complexes with the elements of interest, and the partition coefficient of the complex between the organic solvent and the aqueous solution must be great. The proportion of the volume of the organic phase to those of the aqueous phase represents the enrichment factor. A further advantage is, besides enrichment, that all metal ions not forming complexes with the ligand under the applied conditions remain in the aqueous phase. This is very advantageous for ETAAS, because the major elements of biological materials (Na, K, Ca, Mg), which may give rise to interferences by forming crusts in the graphite furnace, are eliminated.

For calibration one can use aqueous solutions of a salt of the element which have undergone the same treatment as the aqueous solution of the sample. The effectiveness of the procedure is controlled by the use of reference material of the same or similar composition. Interferences are to be expected if the sample contains substances forming complexes more stable than those of the added ligand.

This method has been successfully applied to the determinations of nickel and cobalt in urine and of lead in whole blood. For the determination of cobalt a 4-ml sample of urine is buffered to pH 4.4 and 0.5 ml of hexamethylene ammonium-hexamethylene dithiocarbamidate is added to a mixture of diisopropylketone and xylene. After 5 min of vigorous mechanical shaking, the sample mixture is centrifuged for 15 min at 4000 rpm in order to separate the two phases. The

organic phase containing the cobalt in a concentration eight times
that of the original urine is used for ETAAS determination [34].

6.3. Mineralization

The most frequently applied procedure for the preparation of samples
from biological materials is mineralization. It is an indispensable
reference method in the development of other preparation procedures
when adequate reference materials are not available. It is also
advised for those materials in which the elements cannot be released
from their bondings in the biological matrix by acids or complexing
agents. Many samples of tissue need to be mineralized in order to
transform them into the homogeneous solutions necessary for aliquotation, as well as to eliminate the organic constituents possibly interfering in the determination procedure. Voltammetric methods such as
polarography and cathodic and anodic stripping voltammetry (CSV and
ASV) require complete elimination of the organic constituents, because
many organic substances may also undergo reactions at the electrode
and thus affect the determination of the elements of interest. A
further advantage of mineralization is that in most cases the elements
can be brought to a defined oxidation state.

In recent years several reviews on the mineralization of biological materials have been published [35-37]. An optimal mineralization procedure should fulfill the following requirements:

1. The organic constituents of the matrix must be completely
   eliminated.
2. The content of the elements of interest is not to be
   altered, neither by contaminations nor by losses.
3. The ions of the elements should be completely solubilized
   and be in a chemical form suited for the determination
   procedure.

While the first requirement is easy to accomplish, the two others
are rather difficult or can only incompletely be fulfilled: Losses

will arise from the volatility of elements or their compounds; the formation of insoluble compounds with reagents, the vessel material, or ions inherent to the biological material; and adsorption to the vessel surface or sparingly soluble precipitations of major elements. Contaminations are caused by the material of the vessels and apparatus, the environment in the laboratory, and especially the necessary reagents. Two completely different kinds of mineralization procedures are possible: dry ashing and wet mineralization.

6.3.1. Dry Ashing, Incineration, and Oxygen Combustion

In the classic method of ashing biological material the only reagent used is oxygen from the air. The reaction is exothermic and the temperature within the reaction vessel cannot be controlled. This method of sample preparation is simple and, since large numbers of samples can be handled, often preferred to wet oxidation methods. In general, dry ashing leads to serious losses due to volatilization and adsorption onto the crucible, where the analyte may form a glass or refractory material resistant to solubilization by acids. Contaminations result mainly from the material of the crucible, the furnace, and the air in the laboratory. This method is not suitable for the preparation of samples for trace element analysis.

The main disadvantage of classic incineration is the uncontrolled high temperature; otherwise, oxygen being the only reagent involved, this procedure would be the least contaminating of the two. This problem can be overcome by using activated oxygen [38,39]: Oxygen at reduced pressure (4 Torr) is conducted through a high-frequency electromagnetic field, where molecular oxygen is split into atoms in an excited state. This activated oxygen reacts with the biological material at low temperature (<150°C) and thus prevents the formation of a glass or refractory. The mineralization is carried out in a closed system, oxygen being the only reagent, and therefore contaminations are almost excluded. Losses of volatile elements or compounds may occur owing to the reduced pressure.

The resulting finely divided residue is easily dissolved in dilute mineral acid.

The disadvantages of this method are that only small samples can be mineralized, because they must be spread in a very thin layer to ascertain the reaction with oxygen; the reaction takes a long time (8 hr or more), and the cost of the apparatus is high.

6.3.2. Wet Mineralization

The oxidizing degradation of the biological material using suitable reagents is the mineralizing procedure usually employed. The volatile reaction products of the organic constituents of the matrix and of the reagents can be eliminated, whereas the ions remain in solution. Sometimes this method is used only to solubilize the matrix without completely eliminating the organic constituents, as for subsequent determinations by AAS, for example.

Contaminations in wet mineralization are mainly due to the reagents and the material of the vessels and apparatus. The preferred construction material is silica, since most wet oxidations are done in strongly acidic media.

Losses arise mainly from very volatile elements and compounds, and much less from adsorption and the formation of insoluble compounds.

Three groups of reagents are used in wet mineralization:

1. Oxidants: HNO_3, H_2O_2, $HClO_4$, etc.
2. Solubilizers and neutralizers: H_2SO_4, HCl, NH_3
3. Catalysts: H_2SO_4, Fe^{2+}, Ag^+

The consumption of reagent follows the sequence

oxidants > solubilizers > catalysts

In the choice of a reagent the following factors must be considered:

1. The purity of the reagents and their consumption, for these determine the possible contaminations. The blank values

of the reagents determine the determination limit of the analytical method.

2. The possible formation of interfering substances with constituents of the biological material. For example, materials with a high calcium content should never be mineralized in the presence of H_2SO_4, since the sparingly soluble calcium sulfate formed can cause losses of trace elements by adsorption or coprecipitation.

3. The reagent's suitability for the determination method and the ease with which excess reagent can be removed. For example, H_2SO_4 should not be used for samples to be measured by ETAAS; residual H_2SO_4 produces strong fumes during atomization which disturb the measurement.

4. Possible hazardous reactions. Under certain circumstances the use of concentrated $HClO_4$ as well as H_2O_2 can result in extremely violent explosions.

A major mistake that is often made is to treat the dried biological material with concentrated acids (e.g., HNO_3). The surface of the material becomes then hydrophobic, and the number of reactive sites is diminished, which results in increased consumption of reagent and lengthier mineralization. Therefore, whenever possible, wet mineralization of biological material should start with dilute acid.

A wet mineralization is best performed in two consecutive steps. First, the biological material is hydrolyzed by dilute acid—preferably the same which serves in concentrated form as the oxidant—at low temperature ($<90°C$). Hereby most biological material is dissolved. Depending on the kind of matrix, some organic products of hydrolysis are volatile and disappear; hence these compounds must not be oxidized in the second step and the consumption of oxidant is smaller. An enlarged surface during hydrolysis favors oxidation. In the second step, water is evaporated until the concentration of the oxidant is sufficient to initiate the oxidation of the organic material. If necessary, more concentrated oxidant is added. The oxidation of the matrix can be described by the following equation:

(organic material) (MX) (MY) $\xrightleftharpoons{\text{oxidant}}$ (CO_2) (NO_x) (SO_2) (H_2O) (MX) (MY)

where M stands for cations, X for anions from the reagents, and Y
for anions inherent to the matrix. The thermodynamic equilibrium
will tend to the right owing to elimination of the volatile products
from the reaction mixture. The kinetics of the reaction depend above
all on the applied temperature.

There are three different systems generally used for wet mineralization:

1. The open system without reflux
2. The open system with forced reflux
3. The closed system

In the open system without reflux, usually beakers, dishes, or
round-bottom flasks covered with watch glasses are used; the equilibrium will tend to the right side. The maximum temperature is limited
to the boiling point of the lowest boiling component, which thus
determines the duration of the digestion. The consumption of oxidants
like HNO_3 or H_2O_2 is mainly governed by their losses from distillation. Large samples can be treated in this way and the organic constituents are completely destroyed. There are high risks for contaminations not only from the reagents but also from the environment
in the laboratory, since the temperature at the openings of the vessels is nearly the same as in the reaction mixture and thus strong
air convection arises. Volatile elements like mercury are partially
lost. This procedure is very cheap but not suitable for trace element analysis.

In open systems with forced reflux mainly round-bottom flasks
with attached air coolers are employed; the equilibrium tends also
to the side of the reaction products, but here the maximum temperature is limited by the boiling point of the highest boiling component
of the mixture, because the lower boiling components are refluxed;
the higher temperature shortens the duration of the mineralization
considerably. The consumption of reagents is decreased by more than
one order of magnitude compared with the open system without forced

reflux, and the low consumption of reagents strongly diminishes the risks for contaminations. The attached cooling system eliminates contaminations from the laboratory environment, since the temperature at the openings of the coolers is low. Large samples can be treated with complete elimination of the organic substances. The resulting solutions are very suitable for voltammetric determinations. A partially automated apparatus allowing the simultaneous mineralization of six samples is now commercially available [40,41].

Mercury is mainly lost in the early stage of a wet mineralization. In the presence of great quantities of easily oxidizable organic materials, mercury can be reduced to the element and volatilized even in the presence of oxidizing agents. By the addition of thioacetamide to the dilute acid for hydrolysis, insoluble mercury sulfide is formed and will resist to redissolution till all organic material has been oxidized. Only in the late stage of the mineralization and in the presence of strong oxidants does the mercury sulfide dissolve. Now the mercury ions remain in solution without danger of losses. These solutions are suitable for the voltammetric determination of mercury at the rotating-disc gold electrode, a very sensitive alternative method to the usually applied cold vapor AAS [42].

The closed system for wet mineralization consists of a stainless steel pressure container into which a sealed vessel from suitable non-contaminating material (Teflon or silica) is inserted. After the sample and the necessary oxidant—mostly concentrated HNO_3—are placed into the inner vessel, the pressure container (mineralization bomb) is tightly closed and heated. The system is fitted with a security device to prevent bursting. The volatile reaction products cannot be eliminated from these bombs. Therefore the oxidation of the organic compounds is never complete. Since the vessel is tightly closed from the beginning of the mineralization, the advantages of acid hydrolysis cannot be used. The applicable temperature is limited by the originating pressure, depending on the size and kind of sample. The consumption of reagents is comparable to that of the open system with forced reflux. Contaminations by the laboratory environment are excluded and losses cannot occur.

This last-mentioned kind of wet mineralization is frequently
used for the preparation of small samples for subsequent determinations by AAS, provided that the remaining organic substances do not
distrub the measurements. Solutions prepared by this system are not
suitable for voltammetric measurements. For this purpose the solutions need further oxidation in an open system [43].

REFERENCES

1. I. J. T. Davies, M. Musa, and T. L. Dormandy, *J. Clin. Pathol.*, *21*, 359 (1968).

2. J. Versieck, F. Barbier, R. Cornelis, and J. Hoste, *Talanta*, *29*, 973 (1982).

3. J. Versieck and R. Cornelis, *Anal. Chim. Acta*, *116*, 217 (1980).

4. J. Versieck, *Trends Anal. Chem.*, *2*, 110 (1983).

5. R. E. Thiers, in *Methods of Biochemical Analysis*, Vol. 5 (D. Glick, ed.), Interscience, New York, 1957, p. 274.

6. J. Angerer, K. H. Schaller, and H. Seiler, *Trends Anal. Chem.*, *2*, 257 (1983).

7. J. R. Moody and E. S. Beary, *Talanta*, *29*, 1003 (1982).

8. J. M. Mattinson, *Anal. Chem.*, *44*, 1715 (1972).

9. R. P. Maas and S. A. Dressing, *Anal. Chem.*, *55*, 808 (1983).

10. D. P. H. Laxen and R. M. Harrison, *Anal. Chem.*, *53*, 345 (1981).

11. R. W. Heiden and D. A. Aikens, *Anal. Chem.*, *51*, 151 (1979).

12. J. R. Moody, *Anal. Chem.*, *54*, 1358A (1982).

13. S. A. Katz, *Int. Biotechn. Lab.*, December, 1984, p. 10.

14. B. Sansoni and G. V. Iyengar, in Internat. At. Energy Agency, Vienna, Techn. Rept., *197* (*Element. Anal. Biol. Materials*), 57 (1980).

15. D. Behne, *J. Clin. Chem. Clin. Biochem.*, *19*, 115 (1981).

16. K. Heydorn, E. Damsgard, N. A. Larsen, and B. Nielsen, in *Nuclear Activation Techniques in the Life Sciences*, International Atomic Energy Agency, Vienna, 1978, pp. 129-142.

17. H. Jürgensen and D. Behne, *J. Radioanal. Chem.*, *37*, 375 (1977).

18. M. Stoeppler and H. W. Nürnberg, in *Metalle in der Umwelt* (E. Merian, ed.), Verlag Chemie, Weinheim, 1984, p. 45.

19. G. Hering, G. Lüönd, T. Briellmann, H. Guggenheim, H. Seiler,

and G. Rutishauser, in *Urolithiasis and Related Clinical Research* (P. O. Schwille, L. Smith, W. G. Robertson, and W. Vahlensieck, eds.), Plenum Press, New York, 1985, pp. 205-208.

20. V. Hudnik, M. Marlot-Gomiscek, and S. Gomiscek, *Anal. Chim. Acta, 157*, 303 (1984).

21. G. V. Iyengar, *J. Pathol., 34*, 173 (1981).

22. G. V. Iyengar, K. Kasperek, and L. E. Feinendegen, *Radioanal. Chem., 69*, 463 (1982).

23. M. Stoeppler, H. W. Durbeck, and H. W. Nürnberg, *Talanta, 29*, 983 (1982).

24. R. A. Lewis (ed.), *Richtlinien für den Einsatz einer Umweltprobenbank in der Bundesrepublik Deutschland*, Biogeographisches Institut der Universität des Saarlandes, Saarbrücken, 1985.

25. G. V. Iyengar, K. Kasperek, and L. E. Feinendegen, *Sci. Total Environ., 20*, 1 (1978).

26. G. V. Iyengar, K. Kasperek, and L. E. Feinendegen, *Analyst, 105*, 794 (1980).

27. J. R. Moody, *Trends Anal. Chem., 2*, 116 (1983).

28. K. Heydorn and E. Damsgaard, *Talanta, 29*, 1019 (1982).

29. J. L. M. deBoer and F. J. M. J. Maessen, *Anal. Chim. Acta, 117*, 371 (1980).

30. Richtlinien der Bundesärztekammer zur Durchführung von Massnahmen der statistischen Qualitätskontrolle und von Ringversuchen, *Dtsch. Aerztebl., 68*, 2220 (1971).

31. K. H. Schaller, J. Angerer, G. Lehnert, H. Valentin, and D. Weltle, *Arbeitsmed. Sozialmed. Praeventivmed., 19*, 79 (1984).

32. M. Stoeppler, J. Angerer, M. Fleischer, K. H. Schaller, in *Analyses of Hazardous Substances in Biological Materials*, Vol. 1 (J. Angerer and K. H. Schaller, eds.), Verlag Chemie, Weinheim, 1985, pp. 79-91.

33. M. Stoeppler and K. Brandt, *Fresenius Z. Anal. Chem., 300*, 372 (1980).

34. E. Schumacher-Witthopf and J. Angerer, *Int. Arch. Occup. Environ. Health, 49*, 77 (1981).

35. B. Sansoni and G. V. Iyengar, *Ber. Kernforschungsanlage Juelich, 13* (1978).

36. P. Tschöpel, in *Ullmanns Encyklopädie der technischen Chemie* (W. Foerst, ed.), Verlag Urban und Schwarzenberg, München, Berlin, Vol. 5, 4th ed., 1980, pp. 27-40.

37. H. Seiler, J. Angerer, M. Fleischer, G. Machata, W. Pilz, K. H. Schaller, M. Stoeppler, and H. Zorn, in *Analysen in Biologischen Material*, Vol. 2 (D. Henschler, ed.), Verlag Chemie, Weinheim, 1981, pp. 23-60.

38. J. E. Patterson, *Anal. Chem.*, *51*, 1087 (1979).

39. G. Kaiser, P. Tschöpel, and G. Tölg, *Fresenius Z. Anal. Chem.*, *253*, 177 (1971).

40. H. Seiler, *Labor-Praxis*, *3*, 23 (1979).

41. H. Seiler, D. Schlettwein-Gsell, G. Brubacher, and G. Ritzel, *Mitt. Geb. Lebensmittelunters. Hyg.*, *68*, 213 (1977).

42. M. Leu and H. Seiler, *Fresenius Z. Anal. Chem.*, in press.

43. L. Kotz, G. Henze, G. Kaiser, S. Pahlke, M. Veber, and G. Tölg, *Talanta*, *26*, 681 (1979).

Author Index

Numbers in parentheses are reference numbers and indicate that an author's work is referred to although his name may not be cited in the text. Underlined numbers give the page on which the complete reference is listed.

A

Aaseth, J., 187(120), 199
Abdulla, M., 219(129), 228
Abe, T., 235(21), 253
Abernathy, R. P., 218(121), 228
Abracchio, M. P., 267(18,19), 276, 277; 288(71,73,76), 289 (71,76), 296(127), 300, 303
Adams, F., 145(117), 155
Adams, R. J., 173(68), 197
Agami, M., 76(30), 97
Aggett, P. J., 215(93), 226
Ahrens, L. H., 5(14), 20
Ahrland, S., 6(20), 20; 33(16), 63
Aikens, D. A., 309(11), 334
Akera, T., 173(70), 197
Alessio, L., 242(68), 243(68), 255
Alexander, J., 187(120), 199
Alexander, J. D., 130(32), 151
Alfthan, G., 213(85), 214(85), 226
Allen, K. G. D., 217(118), 227
Ally, A. I., 177(85), 197
Alper, J. C., 210(73), 226
Al-Rabee, R. H., 272(32), 277
Alvarez, A. P., 164(28), 195
Amantini, L., 172(57), 196
Ambrose, A. M., 185(112), 198
Ames, B. N., 286(46,47), 299
Andersen, R. D., 284(37), 298

Anderson, O. E., 146(128), 155
Anderson, P. D., 101(10), 117
Anderson, Y., 264(14), 266(14), 276
Andersson, A., 142(105), 154
Andreae, M. O., 19(28), 20
Andrew, R. W., 101(15), 115(35), 117, 118
Angelosanto, F. A., 251(118), 257
Angerer, J., 308(6), 314(6), 316 (6), 320(6), 323(6), 324(31), 325(31), 326(32), 328(34,37), 334, 335
Anonymous, 203(2), 212(82), 213 (84,87), 214(2), 222, 226
Antonovics, J., 148(145), 156
Arany, J., 238(46), 254
Aranyi, C., 283(16), 298
Archer, P., 292(100), 301
Arlauskas, A., 286(59), 299
Armstrong, M. A., 294(117), 302
Arnon, D. I., 73(15), 97
Arvilommi, H., 213(85), 214(85), 226
Asami, T., 19(28), 20
Au, W., 292(101), 301
Aubert, M., 88(80), 100
Aungst, B. J., 160(3), 194
Auriti, L., 242(69), 243(69,93), 244(93), 255, 256
Awasthi, Y. C., 205(20), 223
Azam, F., 91(83,85), 92(83), 100
Azarnoff, D. L., 172(65), 196
Azhajev, A. A., 241(58), 255

337

Subject Index